T0388341

Nanoscale Semiconductors

This reference text discusses the conduction mechanism, structure construction, operation, performance evaluation, and applications of nanoscale semiconductor materials and devices in very large-scale integration (VLSI) circuit design.

The text explains nanomaterials and devices and analyzes its design parameters to meet the sub-nano-regime challenges for complementary metal–oxide–semiconductor devices. It discusses important topics, including memory design and testing, fin field-effect transistor (FinFET), tunneling field-effect transistor for sensor design, carbon nanotube field-effect transistor (CNTFET) for memory design, nanowire and nanoribbons, nano-devices based on low-power-circuit design, and microelectromechanical systems design.

The book

- discusses nanoscale semiconductor materials, device models, and circuit design.
- covers nanoscale semiconductor device structures and modeling.
- discusses novel nano-semiconductor devices such as FinFET, CNTFET, and nanowire.
- covers power dissipation and reduction techniques.

Discussing innovative nanoscale semiconductor device structures and modeling, this text will be useful for graduate students and academic researchers in diverse areas such as electrical engineering, electronics and communication engineering, nanoscience, and nanotechnology. It covers nano-devices based on a low-power circuit design, nanoscale devices based on digital VLSI circuits, and novel devices based on an analog VLSI circuit design.

Nanoscale Semiconductors

Materials, Devices and Circuits

Edited by
Balwinder Raj and
Ashish Raman

CRC Press
Taylor & Francis Group
Boca Raton London

CRC Press is an imprint of the
Taylor & Francis Group, an **informa** business

First edition published 2023
by CRC Press
6000 Broken Sound Parkway NW, Suite 300, Boca Raton, FL 33487-2742

and by CRC Press
4 Park Square, Milton Park, Abingdon, Oxon, OX14 4RN

CRC Press is an imprint of Taylor & Francis Group, LLC

ISBN: 978-1-032-30754-1 (hbk)
ISBN: 978-1-032-31792-2 (pbk)
ISBN: 978-1-003-31137-9 (ebk)

DOI: 10.1201/9781003311379

Typeset in Sabon
by Apex CoVantage, LLC

Dedicated to my wife Dr Madhu Bala for her constant support during the course of this book

–Balwinder Raj

Dedicated to my wife Ms Deepti for her constant support during the course of this book

–Ashish Raman

Contents

Preface

Today, nanoscale materials and devices have become essential and are increasingly being used for advanced circuit design. It has almost become impossible to imagine a world where people can live without these nanoscale materials, devices and circuits. With the rapid development in the electronics business, a number of inventions have come into view including hardware for communication systems, the Internet of Things (IoT), artificial intelligence (AI), embedded systems and very large-scale integration (VLSI) design used for advanced applications. These are being used by individuals of different age groups and in almost all business processes because of the significant rise in productivity of the work done and overall efficiency.

Chapter 1: Tunneling field-effect transistors (TFETs) show tremendous potential from a low-power application point of view. However, the challenge to gain a high Ion/Ioff ratio in combination with a subthreshold swing <60 mV/decade still persists for devices to be used in high-performance circuits. Achieving this goal of heterostructure TFET devices requires the combined usage of many technologies. In lower technology nodes, its fabrication becomes viable and can be a real, standalone device comparable to metal–oxide–semiconductor field-effect transistors (MOSFETs).

Chapter 2: The fundamentals of TFETs, along with unit configuration, operating theory, and critical electrical properties of TFETs, are discussed. This chapter also investigates the differences between a standard MOSFET and a TFET in terms of operating theory and electrical characteristics. The issues with the TFETs' simple implementation, especially the issue of low ION are highlighted. The techniques used to improve ION in TFETs are also given. Various electrical properties of TFETs are discussed, which have a significant influence on how they are used in different circuits. The use of TFETs in digital circuits, memories and analog circuits is reviewed.

Chapter 3: The discussion starts with the introduction to the necessities of technical transition from poly-Si/SiO2 to high-k/metal gate and the demand for the atomic layer deposition (ALD) of metal gates due to the special challenge of the three-dimensional (3D) devices. Challenges associated with high-k dielectric with polysilicon are also discussed in detail. A review of the ALD of metal gates for both negative–metal–oxide–semiconductor field-effect transistors (NMOSFETs) and p-type metal–oxide–semiconductor field-effect transistors (PMOSFETs) based on state-of-the-art ongoing research is presented.

Chapter 4: This chapter deals with the fabrication process of MBCFETs using a conventional complementary metal–oxide semiconductor (CMOS) process. Because of the vertically stacked double-bridge channels, the MBCFET can drive 4.6 times more current than a normal-plane MOSFET at about the same threshold voltage. Through the implementation of a simple oxidation procedure, the leakage and electrostatic properties are enhanced. The most appealing feature is that it can improve device performance by a simple oxidation process rather than by modifying a complicated photo-patterning process. The suggested approach may greatly reduce the driving current through the core insulator while suppressing leakage current.

Chapter 5: The need for scaling in the semiconductor industry restricts the use of copper (Cu) as an interconnect material at lower technology nodes. At high frequencies and lower technology nodes, Cu interconnects suffer from size effect, electromigration, higher coupling noise, reliability issues, and dispersion effects. Graphene nanoribbons (GNRs) and carbon nanotubes (CNTs) are potential materials to replace Cu for interconnect applications to mitigate all these issues. However, GNRs' and CNTs' higher growth temperature and fabrications challenges may limit the performance of such emerging materials. Therefore, Cu–graphene hybrid interconnects can commendably fulfill these future requirements. Moreover, the use of novel techniques such as graphene-based through-silicon vias (TSVs), ThruChip Interfaces, spintronics and plasmonic and optical-based interconnects can be envisaged for the design and fabrication of ultra-high-performance heterogeneous integrated circuits.

Chapter 6: Graphene nanoribbon field-effect transistors (GNRFETs) dissipate less power than CMOSs, despite the fact that CMOSs are irreplaceable. Beyond 32nm, carbon nanotube field-effect transistors (CNTFETs) dissipate less power than CMOSs. Circuits based on GNRFETs are more powerful than circuits based on CMOSs beyond 32nm. The prospect of GNR in VLSI is intriguing, and it might lead the way for MOSFET technological advancements. As a result, GNR might be used as a CMOS transistor substitute in devices larger than 32nm.

Chapter 7: The rapid advancements of technology have spurred scientific research with enormous prospects, and from this chapter, it can be interpreted that fin field-effect transistors (FinFETs) seem to be a superior alternative compared to the conventional CMOS and other emerging technologies due to its high performance and smaller dimensions. In this chapter, the authors evaluate the electrical characteristic of a FinFET circuit by calculating the ION and IOFF current of N-type FinFET and P-type FinFET in both H℘ and low-standby power (LSTP) model of FinFET from 7nm to 20nm technology. It is observed that if we apply back gate bias from 7nm to 20nm technology, there is no impact on ION and IOFF current both in H℘ and LSTP model. In the LSTP model, the ION is larger than H℘ model because it is a lower standby-power library that suppresses the unwanted leakage current. In the future, FinFET-based leakage reduction methods are preferred for low-power consumption applications in near-term scenario.

Chapter 8: After obtaining all the results we can conclude that operating impact ionization avalanche transit time (IMPATT) diodes at higher frequencies can help reduce the noise measures of the device, thus effectively improving the noise performance of the device. One of the reasons for the reduced noise measure at higher frequencies is the reduction of the depletion width at high frequencies. The reduction in the depletion region also leads to a decrease in ionization and thus may affect the performance of the device. The power-handling capacity of the device is reduced due to the decrease in ionization which is not a favorable property for IMPATT diodes.

Thus, in order to achieve optimum performance from the device, the trade-off between power and noise must be considered carefully. To achieve a good balance between power and noise, the device must be carefully fabricated using proper structural and doping parameters required for the device to perform at that desired operating frequency.

Chapter 9: Various advanced packaging technologies are proposed for microelectromechanical systems (MEMs) and chips (sensors). The same types of challenges are observed in testing packaged MEMS (conventional) and sensors. There is an added complexity when testing a subsystem while employing 2.5D and 3D techniques related to packaging.

Chapter 10: This chapter provides a detailed description of how TFETs work including their advantages, drawbacks, existing solutions, and applications. TFETs, which work on the inter-band tunneling principle and are not restricted by thermionic emission, have very low OFF current and subthreshold swing as compared to CMOSs. However, TFETs have a comparatively low ON current and high ambipolarity than metal–oxide semiconductors (MOSs), which restricts their use in circuits. Besides, researchers had suggested many structural and physical modifications to upgrade TFET characteristics. In this sequence, the authors explore a hetero-material hetero-dielectric remote field electromagnetic technique (HM-HD-RFET) in a configurable physical unclonable function (PUF) architecture. Basically, an RFET is an electrically doped device that can be reconfigured either into n or p TFET or n or p MOSFET based on the potential applied over external drain and source electrodes. Indeed, Si-based RFET produces satisfactory MOSFET results just like conventional one but lacks ON current and higher ambipolarity as in Si-TFET. Therefore, HM-HD-RFET proposed in this chapter produces sufficient drain current in all FETs that are capable of driving circuit applications. Also, in this chapter, the reconfigurable nature of RFET is explored in the mixed-signal-based hardware security application through the construction of a configurable ring oscillator (CRO)-PUF architecture that produces 2N secure keys as compared to 2N keys as in conventional RO-PUF. Hence, this chapter opens new avenues to investigate TFET and explore its utilization in novel applications.

Chapter 11: Junctionless transistors (JLTs), at the nanoscale regime, give a good immunity to the short channel effects and produce a higher Ion/Ioff ratio. Also, the involvement of the high-k spacer helps to massively reduce the leakage current. Although the band-to-band tunneling produces the minority charge carriers in the transistors, it is suppressed within the channel itself in the case of JLTs. The reduction in oxide thickness makes a JLT superior to a conventional FET in terms of speed. Hence, it may be considered as the potential alternative for next generation ultra-low-power VLSI design. The junction isolation in bipolar JLTs helps to completely deplete the channel in the OFF state.

Since JLTs are of narrow width, strained silicon technology may be used to configure the device to enhance performance. The work function engineering may also be implemented to the gate electrode for the maintenance of the threshold voltage.

Acknowledgment

We, the editors, would like to thank everyone who participated in finalizing this book. Many people have contributed greatly to this book, *Nanoscale Semiconductors: Materials, Devices and Circuits*. We would like to acknowledge all of them for their valuable help and generous ideas in improving the quality of this book. With our feelings of gratitude, we would like to introduce them in turn. The first mention is to the authors and reviewers of each chapter. Without their outstanding expertise, constructive reviews and devoted effort, this comprehensive book would not have been possible. The second mention is to the CRC Press/Taylor & Francis Group staff, especially Gauravjeet Singh Reen, Isha Ahuja and their team for their constant encouragement, continuous assistance and untiring support. Without their technical support, this book would not be completed. The third mention is to the editors' families for being the source of continuous love, unconditional support and prayers not only for this work but throughout our lives as well. Last but far from least, we express our heartfelt thanks to the Almighty for bestowing over us the courage to face the complexities of life and complete this work.

Balwinder Raj
Ashish Raman

Editors

Dr. Balwinder Raj (MIEEE 2006) completed his B.Tech—Electronics Engineering (PTU Jalandhar), M.Tech—Microelectronics (PU Chandigarh) and PhD—VLSI Design (IIT Roorkee, India) in 2004, 2006 and 2010, respectively. For further research work, European Commission awarded him the Erasmus Mundus Mobility of Life Research fellowship for postdoc research work at the University of Rome, Tor Vergata, Italy, in 2010–2011. Dr. Raj received India4EU (India for European Commission) Fellowship and worked as a visiting researcher at KTH University, Stockholm, Sweden, in October and November 2013. He also visited Aalto University Finland as a visiting researcher during June 2017.

Currently, he is working as Associate Professor at the National Institute of Technical Teachers Training and Research Chandigarh, India since December 2019. Previously, he worked at the National Institute of Technology (NIT Jalandhar), Punjab, India from May 2012 to December 2019. Dr. Raj also worked as Assistant Professor at ABV-IIITM Gwalior (an autonomous institute established by the Ministry of Human Resource Development, Government of India) from July 2011 to April 2012. He received Best Teacher Award from Indian Society for Technical Education (ISTE) New Delhi on July 26, 2013. Dr. Raj received Young Scientist Award from Punjab Academy of Sciences during the 18th Punjab Science Congress held on February 9, 2015. He has also received a research paper award at the International Conference on Electrical and Electronics Engineering held at Pattaya, Thailand, on July 11–12, 2015. Dr. Raj has authored/coauthored three books, eight book chapters and more than 70 research papers in peer-reviewed international/national journals and conferences. His areas of interest in research are classical/non classical nanoscale semiconductor device modeling; nanoelectronics and their applications in hardware security, sensors and circuit design, fin field-effect transistor–based memory design, low-power very large-scale integration (VLSI) design, digital/analog VLSI design and Field Programmable Gate Array (FPGA) implementation.

Dr Ashish Raman completed a BE—Electronics and Communication Engineering, an M.Tech—Microelectronics and VLSI Design (Shri G S Institute of Technology and Science, Indore) and a PhD—VLSI Design (NIT Jalandhar), India in 2003, 2005 and 2015, respectively.

Currently, he is working as Assistant Professor at the National Institute of Technology, Jalandhar, India since September 2007. Previously, he worked at the National Institute of Technology (NIT Jalandhar), Durgapur, India, from January 2007 to August 2007.

Dr. Raman has authored/co-authored one book, five book chapters and more than 50 research papers in peer-reviewed international/national journals and conferences. His areas of interest in research are very large-scale integration (VLSI) circuit design, nanoscale semiconductor devices; nanoelectronics, modeling, low-power VLSI design, digital/analog VLSI design and FPGA implementation and sensor and circuit applications. He is working as a principal investigator of various funded projects including SMDP-C2SD sponsored by MeitY; FIST sponsored by DST; FPGA-based high-speed CCSDS processor for base band receiver, funded by ISRO Banglore; and many more projects. Dr. Raman is a member of the IEEE Electron Devices Society, the IEEE Solid-State Circuits Society and the Institution of Engineers Society, India.

Contributors

Gaurav Aggarwal
RKGEC, Hapur

Shamshad Alam
NIT Jalandhar

Manisha Bharti
NIT Delhi

Ajay Kumar Dadoria
MNIT Bhopal

R. Dhar
IIEST Shibpur

Tanvika Garg
NIT Delhi

Neha Gupta
IISER

Kavita Khare
MNIT Bhopal

Afreen Khursheed
MNIT Bhopal

A. Krishna Kumar
Chaitanya Bharathi Institute of Technology
Gandipet, Hyderabad, Telangana

Somesh Kumar
IIITM Gwalior

Vikas Maheshwari
Guru Nanek Institutions

Md Rashid Mahmood
Guru Nanek Institutions

Alak Majumder
National Institute of Technology
Arunachal Pradesh, Yupia

Manoj Kumar Majumder
IIITM Gwalior

M. Mitra
IIEST Shibpur

Sangeeta Jana Mukhopadhyay
IIEST Shibpur

Tika Ram Pokhrel
National Institute of Technology
Arunachal Pradesh, Yupia

Balwinder Raj
NIT Jalandhar

G. Boopathi Raja
Velalar College of Engineering and Technology
Erode

Chithraja Rajan
V. Ramakrishna

Sri Padmavati Mahila Visvavidyalayam (Women's University)
Tirupati, Andhra Pradesh

Ashish Raman
NIT Jalandhar

Dip Prakash Samajdar
Ajeet Singh
RKGEC, Hapur

Amarah Zahra
NIT Jalandhar

Chapter 1

TFETs, the Nonconventional Transistor Basics

Gaurav Aggarwal and Ajeet Singh

CONTENTS

Tunnel transistors were developed in 1992 by T. Baba; it is a promising candidate for designing very low-power-specific analog integrated circuits (ICs) with high stability over a wide temperature range. As a result, the huge circuits, which are often power-hungry, can be made more energy-efficient, and tunneling field-effect transistors (TFETs) are also recognized as "green transistors." The aim of the chapter is to familiarize readers with TFETs and get an insight into their workings. Throughout this chapter, we describe the basic TFET device structure, understand the basic functionality qualitatively and learn to simulate the basic device structure using TCAD ATLAS software.

We will see that the TFET structure looks symmetrical with respect to the doping concentrations of source and drain, but it exhibits the ambipolar

DOI: 10.1201/9781003311379-1

behavior as the doping type used on the drain side is opposite to that used on the source side. So, when the electrons are the major contributors to the current flow, the TFET behaves as an n-type, whereas p-type behavior is seen when holes play a role as the majority of current carriers. But for a lower Off-state current, this ambipolar behavior needs to be removed, which is done through a heterojunction structure or by introducing some kind of asymmetry into the structure. It is very challenging to realize the nonsymmetric TFETs due to the requirement of high-quality III–V semiconductor materials for reliable production. The device performance degrades due to the non-idealities of crystal structure, process defects, and implant non-uniformities in the actual fabrication that generates the unwanted electron–hole defect states in the band gap. The lifetime of minority carrier and trap states must be considered for device calibration and to achieve the best fit models. The quality of models is generally damaged due to the improper accounting of the intrinsic electron trap states, which can be approximated more correctly by accounting for the additional generation-recombination models.

Finally, a device performance comparison for three kinds of devices—silicon-on-insulator (SOI) metal–oxide–semiconductor field-effect transistor (MOSFET), double-gate TFET (DG-TFET), and SOI TFET—and the ways to improve the ON current of tunnel FETs by using lower band-gap semiconductor material for fabrication in place of silicon is described. The reason for concern with the TFETS is the lower value of ION with respect to MOS devices, making them unsuitable for high-power applications. Also, due to asymmetry of structure, in TFETs, unlike MOSFETs, the source and drain are not interchangeable, resulting in a unidirectional flow of current. This asymmetricity of TFET has its consequences on the circuit density, since the conventional MOSFET circuit layout is density efficient. Mostly for the circuits with bidirectional current conduction mechanism, it is then proposed to implement an alternative topologies approach that can help in gaining more battery life but at the cost of losing the area. When it comes to die size and cost, it becomes difficult to choose between MOSFETs and TFETs. For applications that can be parallelized, there exists a trade-off between the core voltages and the core gate count. If we want improved performance per unit wattage then the circuit needs to be operated at low voltage and low frequency, leading to a low performance per unit area, that is, a greater number of cores are needed to achieve the similar performance of the circuit. In many applications, in circuits that do not use the concept of parallelism, the customized use of complementary metal–oxide semiconductor (CMOS) cores at high frequencies with low-power TFET cores at low frequencies is visualized.

This chapter also explores the critical device performance parameters for such devices, how to analyze the simulation results and a brief discussion on the future scope of such devices. Improved versions for these devices can be better designed if we clearly understand the parameters, which can be tuned either through the device design, through fabrication or by using new materials. It is also very important to include the right models in device simulation to further match it with practical results. The fabrication node or the technology chosen for practical applications plays a major role in device performance enhancements. The undesirable gate or source to drain leakages are the challenges, to be addressed before the commercial acceptance of the device in the industry. However, these novel devices may become an essential part of circuit design if we can reduce the complexities in the co-fabrication of these devices, along with conventional FETs, to gain the optimum functionality with minimum power consumption and, thus, will be an added advantage in the digital world.

1.1 INTRODUCTION

In the last two decades, CMOS device technology has undergone a revolutionary change, and the process, as well as design capabilities, have been explored and utilized up to its limits. Integrated circuits are constantly achieving higher performances and are able to embed more functionality per unit area due to the scaled power consumption of these CMOS-based circuits. The wide use of CMOS technology is possible only due to its scalability to smaller dimensions and it has helped in achieving the higher performances, along with a decrease in the power consumption due to voltage scaling. Today's processor performance enhancement has been driven by the scaling of the transistor (MOSFET) technology, which resulted in decreased power consumption while gaining speed. This continued scaling of transistors and the huge rise in density have significantly increased the power density on the chip. The reduction of supply voltages, as Vdd is scaled, has quadratically reduced the power consumption in the dynamic state, but it has greatly impacted the application with static power constraints posed by high-speed computing chips' requirement of energy efficiency. As we see, to get a similar drive current for maintaining the drive strength, the Vth (threshold voltage) is to be scaled in proportion with the Vdd reduction of a transistor, which causes the off-state leakage to rise exponentially, leading to static leakage and thus the static power. The basics of TFETs state as a test transistor that uses quantum tunneling modulation as the mode of current generation, unlike the standard MOSFETs that use thermionic emission, thus giving a lower value of the subthreshold swing. Therefore, it is a promising candidate for designing analog or mixed-signal integrated circuits, especially when designing for ultra-low power consumption and stability over a wide range of temperatures for a longer life. We discuss the basic TFET device structure, understand the functionality qualitatively, and provide simulations of the simple device structure using TCAD ATLAS software.

Also, real MOSFET behavior is not matched with an ideal switch because of the presence of nonzero off-state leakage, causing power dissipation even in the off condition. Moreover, the MOSFET follows a finite slope due to its non-ideal switching behavior to switch from Off state to ON state. This behavior is simply defined by the parameter called subthreshold swing which represents the change in voltage required to bring one order change in the current. Its mathematical formula is

$$\text{Subthreshold Swing} = \left(\frac{d\,log\left(I_{DS}\right)}{d\left(V_{GS}\right)}\right)^{-1}.$$

The subthreshold swing for a MOSFET is given as

$$\text{Subthreshold Swing} = \left(1 + \frac{C_{DEP}}{C_{OX}}\right)\frac{KT}{q}\ln(10) \geq \frac{KT}{q}\ln(10).$$

Here, C_{DEP} and C_{OX} represent the depletion region and gate oxide capacitances, respectively. In this equation, the first term represents the efficiency with which the gate couples the with applied gate voltage to the potential in the channel region of the MOSFET and is always >1 for typical operating conditions of the MOSFET. The second term represents the value of Boltzmann thermal distribution followed by mobile charge carriers and is limited by constants kT/q to 60mV/decade at room temperature, which implies that even for the perfect gate coupling to the channel potential ($C_{OX} >> C_{DEP}$) making $1 + C_{DEP}/C_{OX}$ as 1, a minimum of 60mV gate voltage is needed to bring one order magnitude change in the current. It can be easily inferred that the fundamental limitation in the switching behavior is kT/q (60mV/decade) at room temperature, and it cannot be reduced below this value as the Boltzmann statistics is being followed for the carrier injection phenomenon. Thus, we need a switch that follows a different mechanism for charge carrier injection and requires vital explorations in the device design. It is observed that TFETs can be used as an alternative to the current device in use and may help in overcoming the limit posed in the subthreshold swing, thus improving the static power dissipation of CMOS devices. The TFET exploits the band-to-band tunneling (BTBT) phenomenon for carrier injection, and this helps lower the subthreshold swing <60mV/decade due to the absence of thermal (kT/q) dependence. TFETs were developed by T. Baba in 1992 and provide an alternative to a conventional switch based on various performance parameters:

- Potential for exceeding the subthreshold swing limit of 60mV/decade
- Ultra-low voltage and ultra-low power
- Higher ION/IOFF current ratio

- Immune to short channel effects
- Highly reduced leakage currents
- Overcoming the speed limitations by utilizing a tunneling mechanism
- An available device operation range for the subthreshold and super-threshold regions
- Fabrication process similar to conventional MOSFETs

These devices can also be utilized for biosensors as alternate methods are being explored in biomolecules detection for comparison and analysis of sensitivity to various devices. These devices, due to their high detection ability, will help improve the quality of medical tests and methods that can be utilized for mass usage. Moreover, the demand for low-power devices in the medical field is encouraging biomedical devices to be designed using new technologies, which can reduce the burden of cumbersome equipment, with smart digital devices replacing them.

1.2 BASICS OF TFETS

1.2.1 TFETs' Basic Definition and Their Workings

TFETs are still treated as an unconventional type of transistor. Figure 1.1 shows the TFET structure on SOI. It can be described as a gated p-i-*n*+ diode, with the gate residing on undoped orintrinsic region, which forms the channel. The *p*+ region is named as the source, and the *n*+region is the drain. A look at the structure gives the impression of its similarity to the conventional MOSFET structure, but there exists a fundamental difference in the switching mechanism. Both these things make this device a prime candidate for low-power circuits. The TFET, as indicated by the name itself, modulates the quantum tunneling through the barrier rather than the thermionic emission over the barrier as in a conventional transistor. This phenomenon is

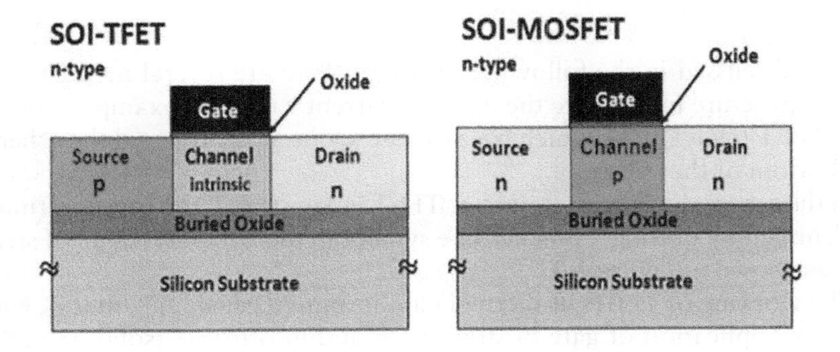

Figure 1.1 Basic structure of a TFET and MOSFET.

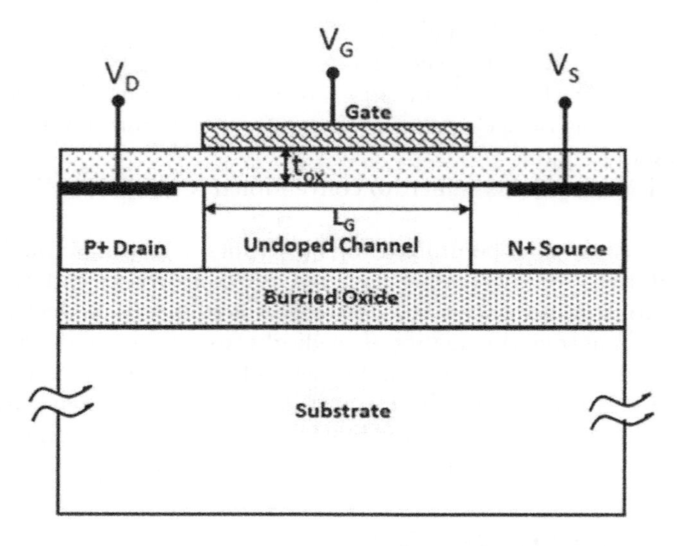

Figure 1.2 Schematic diagram of single gate n-type TFET on SOI substrate.

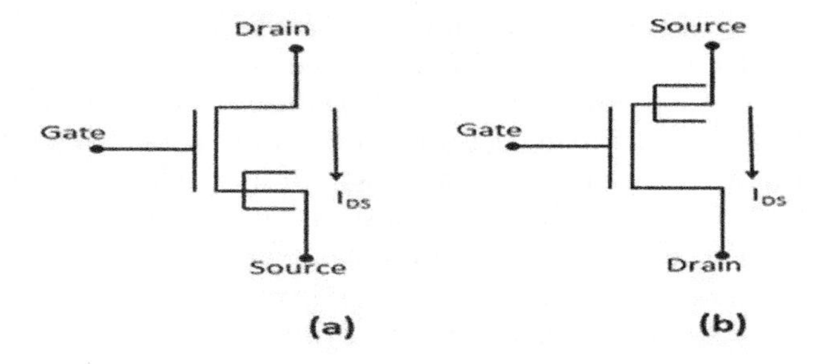

Figure 1.3 Circuit symbol of a TFET: (a) n-type and (b) p-type.

further discussed in the following sections. There are several modifications to this structure to improve the on-state current value, for example, double-gate TFET (DG-TFET), which has two gates, one at the top and the other at the bottom of the substrate.

In the schematic representation of TFET in Figure 1.2, the source terminal is identified by putting a bracket-like symbol in the MOSFET symbol structure as shown in Figure 1.3.

The working of TFETs at thermal equilibrium (dashed line), that is, without any application of gate or drain bias, and in off-state (solid line), that is, only drain bias is applied, as shown in the energy band diagram in Figure 1.4. Since the channel conduction band, *ECchannel* is not in alignment

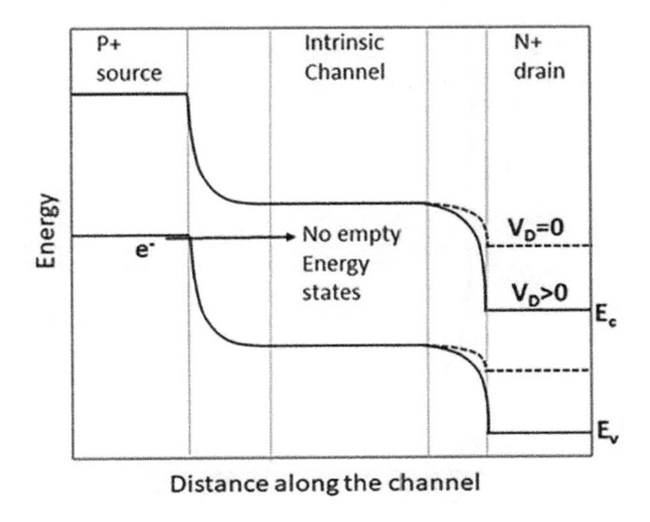

Figure 1.4 Band diagram along the surface of a TFET in the OFF-state and at thermal equilibrium.

with the source valence band, *EVsource*, the electrons cannot tunnel. Also, the probability of tunneling of minority carriers in p-source, that is, the electrons, from *EVsource* to *ECdrain* is almost nil or nonsignificant, so the off-state current is very small. The operating principle of TFET can be understood by qualitatively deriving its transfer and output characteristics. These characteristics are theoretically simulated ones, and their physical significance is discussed in the next section.

1.2.2 Transfer Characteristics

Transfer characteristics refer to Drain Current (ID) versus Gate Voltage (VG) graph for several different values of voltage on the drain side. For an n-type TFET, assume some constant value of Drain Voltage (VD) >0. When VG is zero, the device is in the off-state. As VG is increased, the channel bands shift downward. There would be very low current until the *EVsource* and the *ECchannel* are aligned (Figure 1.5a). As soon as they are aligned, the electrons from the source would tunnel to the channel and drift down to the drain side. This causes a sudden jump in the drain current and a very steep subthreshold slope in TFETs. When VG is further increased, the channel bands shift farther down (Figure 1.5b), thus causing the tunnel width to decrease and, in turn, increasing the number of electrons in the valence band of the source that are able to tunnel, leaving more empty energy states in channel conduction band. Hence, the drain current increases further.

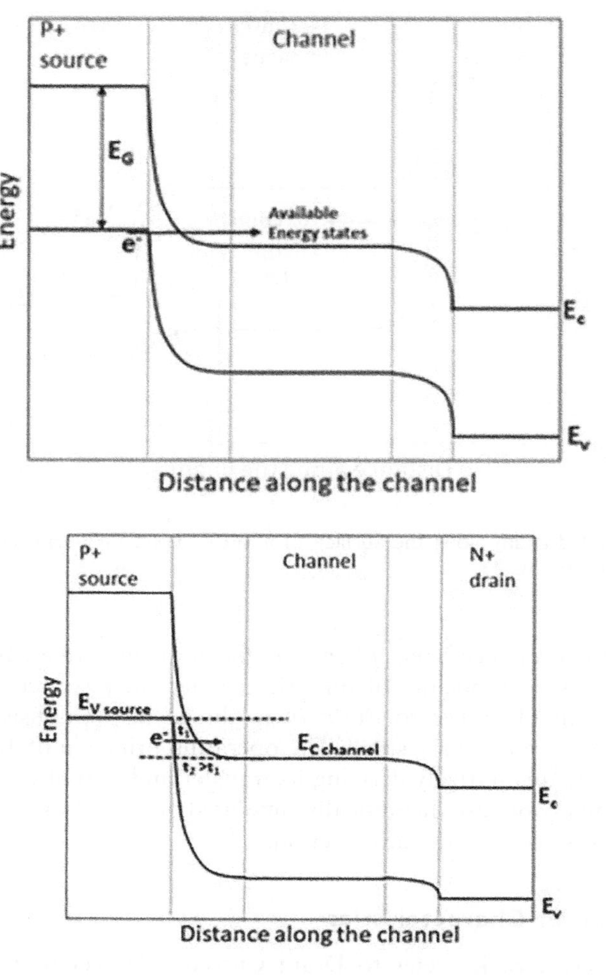

Figure 1.5 Band diagram along the surface of a TFET (a) at the beginning of the ON-state and (b) deep into the ON-state.

It should be noted that the probability of tunneling increases exponentially with a decrease in the tunneling width. As the gate bias is further increased, there comes a situation (around when VG ~VD) when the *ECchannel* is aligned with the *ECdrain*, leading to channel potential pinning with that of drain. This situation happens when the inversion charge concentration becomes equal to the n-drain electron concentration, leading to a short-circuiting of the channel and the drain. In pinning conditions, the drain would try to keep the channel potential at the applied VD; hence, any further increase in VG will not significantly lower the channel bands.

But the electric field across the source–channel junction increases, leading to further thinning of tunneling length and hence an increase in the drain current. As can be seen from the plot in Figure 1.6, the current increase after pinning is at a slower rate as compared to that in the subthreshold region. This is because although the tunneling barrier is thinning, no new empty energy states in the channel or filled states in the source are made available for tunneling (as there is no movement of channel bands due to pinning). Now, for different values of VD >0, the ID–VG curves for VG >0 are shown in Figure 1.7. As expected, for lower drain voltages, the pinning will happen

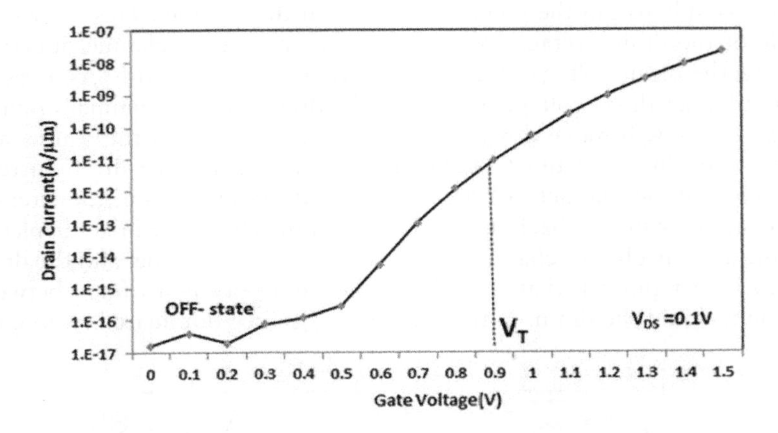

Figure 1.6 Transfer characteristics (ID–VGS) of an n-channel TFET for a positive value of VDS, showing different regions of operations.

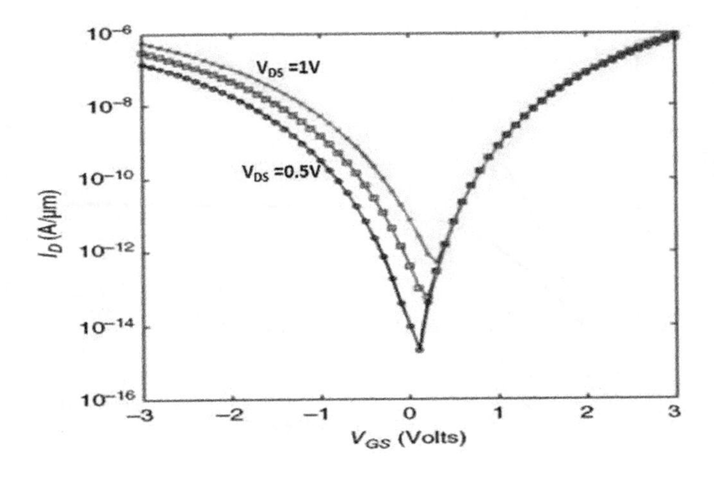

Figure 1.7 Transfer characteristics of TFET.

at lower gate voltage. As can be observed in Figure 1.7, the subthreshold region curves almost overlap for all three values of VD.

1.2.3 Output Characteristics

Output characteristics (Figure 1.8) refer to ID versus VD graph for several different values of voltage on the gate. Considering an n-type TFET, when the drain is at a low voltage and gate voltage is applied, then a channel under the gate is formed having high electron concentrations. Due to the high concentration of electrons, the channel resistance is low, and the drain voltage is able to pin the conduction band in the channel. This causes the whole of the drain voltage to appear across the source–channel junction. So now, the drain voltage modulates the tunnel barrier width. As a result, when a higher drain voltage is applied, the drain and the channel's pinned energy bands will move down, availing more electrons in the source and empty states in the channel for tunneling, hence increasing drain current. Since most of the channel electrons are injected from the drain, an increased drain voltage will pull back its electrons, and the channel will get depleted, causing a nonnegligible channel resistance; hence, the channel and the drain are no longer pinned; that is, a depletion region begins to form between the channel and the drain. In this channel-resistance-dominated region, the

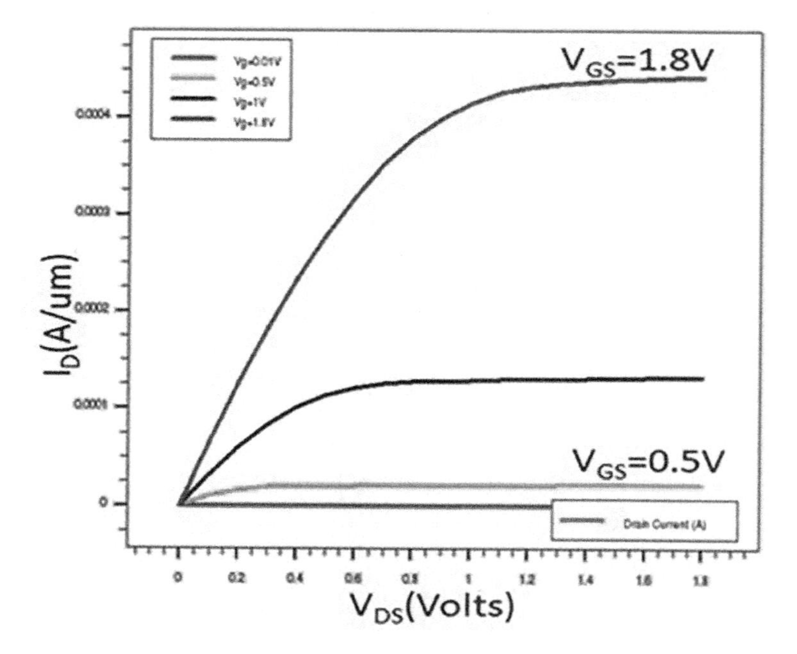

Figure 1.8 Output Characteristics for TFETs.

drain voltage has control over the source—channel tunneling reduces, causing the drain current to increase at a slower rate. As the drain voltage is further increased, at a certain voltage, when the depletion width is equal to that thermal equilibrium, the channel is depleted such that it disappears from the source. This phenomenon is similar to pinch-off in MOSFET. Due to the sufficiently high channel resistance, the lateral electric field from the drain region cannot cross through the channel region, any further increase in drain voltage will no longer affect the drain current. The potential available at the source–channel junction is being controlled primarily by the gate and no longer affected by the drain voltage. Any additional applied voltage on the drain side appears across the drain–channel junction, causing the drain energy bands to move down with respect to the channel.

1.2.4 Ambipolar Behavior of TFETs

The TFET structure represents two junctions for tunneling: one is at the source side and the other one is at the drain side. The source–channel and drain–channel junctions both use the same tunneling mechanism, that is, BTBT, but the BTBT in the drain–channel junction is reduced by lowering the doping levels of the drain or by using other techniques. However, when the same doping concentration but of the opposite type is implemented for the source and drain junctions, then the TFET structure is symmetric, and its behavior is ambipolar in nature. The ambipolar behavior is defined as when both electrons and holes can play a role in the current transport mechanism, with only one being the major contributor in the current transport. When electrons are the major contributors, a n-type behavior is observed, and a p-type for holes contributes to the majority of the same voltage at the drain side, depending on the gate bias. TFETs show similar characteristics for VG 0 for constant positive drain bias. As mentioned, when VG = 0, the device remains in an off state with very low off-state leakage current. When VG is decreased, that is, now is VG > 0, there is negligible current until tunneling starts. Here, as more negative gate bias is applied, there would be a condition in which the *EVchannel* and the *ECdrain* will get aligned, leading to the tunneling of electrons. The holes from the channel can then drift to the source, leading to a drain current that is in the same direction as if the gate is biased positively. As more and more negative bias is applied to the gate, the *EVchannel* and the *EVsource* get aligned, effectively shorting the channel to p-type source. That is the channel is now pinned to the source. Then a further application of negative gate bias will lead to a slower rate of increase in the current. For different values of VD > 0, the ID–VG curves for VG > 0, an increase in the drain voltage will increase the tunneling probability from the channel to the drain. The drain energy bands will be lower for higher drain voltages, leading to an early onset of tunneling. Ambipolar current arises due to undesirable conduction, and it gives rise to higher off-state leakage currents in TFETs. The ambipolar current increases the off-state leakage in particular in TFETs fabricated with small band-gap materials. The ambipolar current that arises due to the BTBT at the drain junction and adds to the off-state leakage may outnumber the leakage currents contributed thermally. Thus, the resultant leakage current is dominantly an ambipolar current rather than thermal leakage. So, as far as possible, the undesirable ambipolar current should be minimized or, if possible, completely cut out to achieve the desired device characteristics. There are ways to evidently reduce this leakage, such as reducing drain doping; separating the drain and the gate, representing an underlap condition of the drain to the gate; creating hetero-junctions; or introducing any kind of asymmetricity in the basic structure. For the simulations done in the next section, a higher doping concentration in the source helps suppress this ambipolar current.

1.3 SIMULATION AND DEVICE MODELING METHODOLOGY

Simulations are done using Silvaco's TCADATLAS software for three devices, that is, SOI-TFET, DG-TFET, and SOI-MOSFET. Their characteristic curves are plotted and compared. Evaluation parameters for comparison of these devices are ION/IOFF, subthreshold swing, and leakage current. The simulation employs a nonlocal tunneling model. A band-gap narrowing effect is included. The effect of concentration-dependent mobility and The Scale-Resolving Hybrid (SRH) models should also be included. The band diagram across a horizontal cutline at off-state is shown in Figure 1.9. The valence band of the source and the conduction band of the channel are not aligned initially. But as a higher gate bias is applied, the simulated band diagrams start modifying the barrier thickness. The thinning of the channeling width as more gate bias is applied can be observed in Figures 1.9 through 1.12. The pinning phenomenon is also evident in Figure 1.12. The simulated transfer characteristics show the expected behavior. The curves of different drain voltages, for negative gate bias sweep, don't overlap as the drain bias directly influences the tunneling current (channel–drain is the tunneling junction), as discussed in the previous section. On the other hand, when the gate bias is swept positively, the drain bias has no direct effect on the tunneling current (source–channel is the tunneling junction); hence, the curves almost overlap for most of the gate bias sweep. As discussed previously, the only effect of drain bias is when the pinning of bands begins. The heterojunction TFET, such as GaSb/InAs TFET, and many other materials chosen for TFET, such as germanium- and indium-based compounds Ge, GeSn, InAs, and InGaAs, are difficult to model and is a limitation to find the optimum TFET material. The non-idealities arising due to trap sites or any defect resulting in non-abrupt band edges play a crucial role and need to be considered in simulations to match the experimental results with theoretical data.

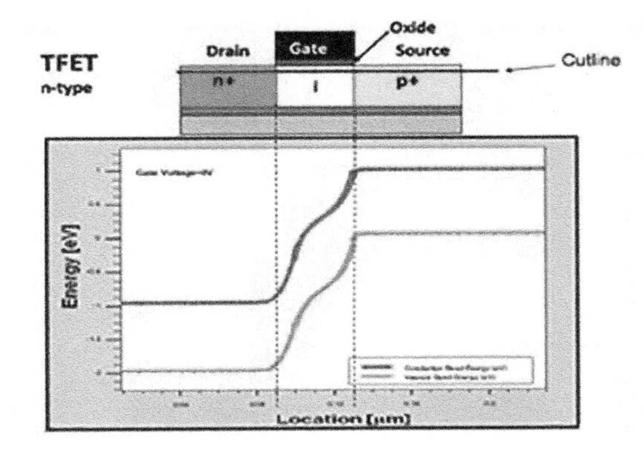

Figure 1.9 Band diagram of off-state TFET across horizontal cutline.

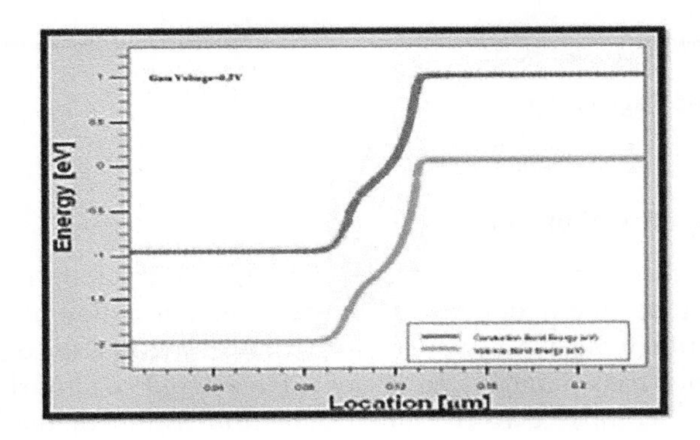

Figure 1.10 Band diagram at gate bias = 0.5V.

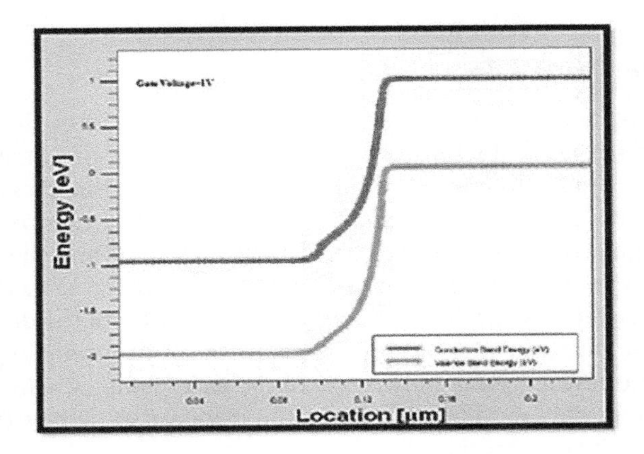

Figure 1.11 Band diagram at gate bias =1V.

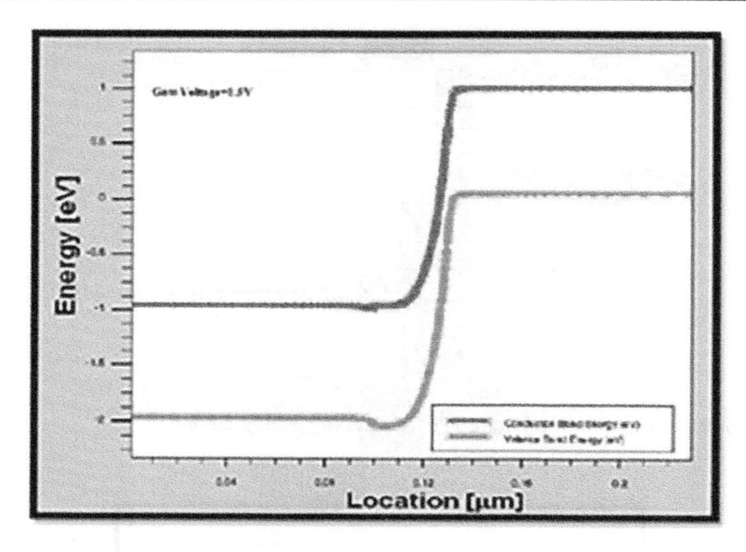

Figure 1.12 Band diagram at gate bias =1.5V.

1.3.1 Impact of Doping

The source and drain doping also play an important role in these devices. When the source side is doped highly, the depletion region at the source–channel junction will have a smaller width, thus shortening the distance traveled by the electrons tunneling from the valance band edge of the source to the channel as compared to the larger depletion width for the low source doping. Therefore, in the case of higher doping of the source, a higher tunneling probability is expected across the source–channel junction, thereby increasing the ON current, resulting in higher drive strength.

1.3.2 Gate Oxide Effects

The oxide capacitance is taken as $C_{OX} = \varepsilon_{OX}/t_{OX}$, where εox is the oxide permittivity and tox is the thickness of gate oxide. We can approximate the gate oxide effects using this simple formula; the gate oxide thickness reduction will increase the gate oxide capacitance. The higher the gate oxide capacitance, the higher the charge in the channel region will be and the better gate control over the channel. The coupling efficiency of the gate to the channel can be increased by maintaining the thinner gate oxides in the device. But at the same time, the thin gate oxides are prone to carrier tunneling from the channel to the gate, causing an increase in the gate leakages, which is highly unacceptable in devices. The end use of the device also defines the leakage limits that can be compromised with gate control.

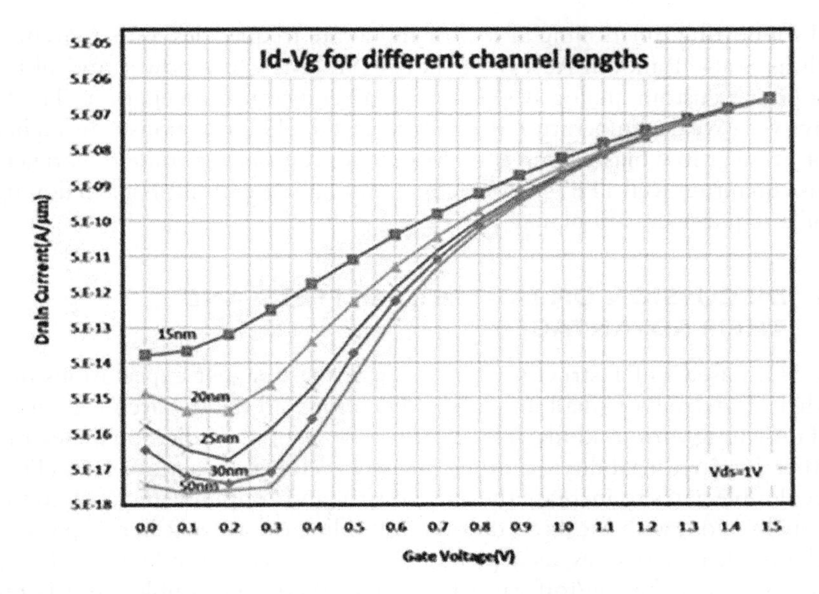

Figure 1.13 Id–Vg plot for different channel lengths of TFET.

1.3.3 Channel Length Affecting Device Parameters

It is very well known that conventional MOSFETs' drain current is largely dependent on the channel length, but interestingly, TFETs' drain current shows little dependence on the length of the channel. The current conduction is largely through BTBT; the tunneling current is governed mainly by the electric field at the source–channel junction and the alignment of the bands on the source side, so a decrease in the channel length will not alter the tunneling current much, as depicted in Figure 1.13. It must be noted that beyond the critical length of the channel, a decrease in the channel length is affected by the direct source to drain leakage.

1.4 APPLICATIONS OF TFETS

TFETs can be a possible replacement for our conventional MOSFETs. As in MOSFETs, BTBT is mostly considered an undesirable secondary effect of short channel lengths. TFETs have been shown to be favorable mechanisms for current injection, which makes further reducing operating voltages a possibility that is not achievable in conventional MOSFET technology. It can give parallelized products, such as graphics cores, giving more battery life at the cost of slightly more chip area.

These new devices are best suited in the applications employing very low standby leakage and low-power circuits because of their loweroff currents and supporting functioning at low or moderate frequencies. Other circuits, such as Static Random Access Memorys (SRAMs), low power and ultra-low-power custom analog ICs with higher temperature strength as TFETs show very weak temperature-dependence, and TFETs demonstrate higher shot noise at low biasing currents. Also, it is proposed by some researchers to use quantum well TFETs to overcome the challenge of ultra-sharp doping profile requirements of planar TFETs.

1.5 DISCUSSION ON DG-TFET, SOI TFETS, AND THEIR ADVANTAGES

The SOI-based TFET structures are not meeting designer expectations due to their drawback of small ion values. Therefore, all the research is focused on fabricating various kinds of structures to overcome smaller ion issues and further improve usability. Another newly developed structure, DG-TFET, have double gates to increase the gate control and coupling efficiency to the channel region, with the expectation of doubling the current in comparison to the single-gate devices, as depicted by the cross section in Figure 1.14.

As expected, the Ion/Ioff ratio for TFETs (SOI and double gate) is far better than in SOI MOSFETs. The subthreshold swing for DG-TFETs is also lower than 60mV/decade. Also, it can be observed that the leakage current is approximately 9 orders lower for both TFET devices as compared to SOIMOSFETs. It can also be concluded that DG-TFETs clearly outperform SOI TFETs in all the evaluation parameters. The reason of concern with TFETS is the lower value of ION with respect to MOS devices, making them

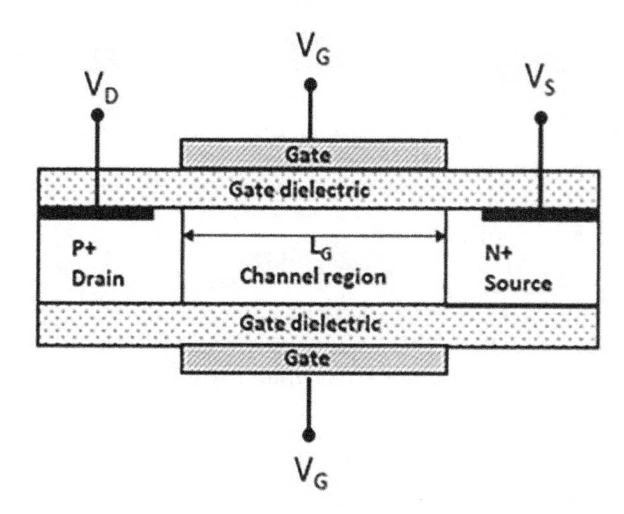

Figure 1.14 Schematic cross-sectional view of a DG-TFET.

unsuitable for high-power applications. Also, due to asymmetry of structure, in TFETs, unlike MOSFETs, the source and the drain are not interchangeable, resulting in a unidirectional flow of current.

The SOI-based structure of TFETs, the SOI-based structure of MOSFETs, and the DG-TFET structures when simulated show a subthreshold swing lower than 60mV/decade for both kinds of TFET structures. Furthermore, a higher amount of on-current was observed in the simulated DG-TFET as compared to single-gate TFET SOI-based structure.

1.6 FABRICATION TECHNOLOGY AND ITS CONSTRAINTS

The challenge behind the actual application is mainly governed by the ease of the fabrication process and the commercial technology that can be sustainably and profitably used to manufacture the devices at low cost. The high-grade raw material should not be a constraint for carrying out volume production. In today's world, lower node technologies are being exploited to fabricate TFETs, but the device parameters, if needed to tune for various applications, put a challenging task ahead for the technocrats to keep pace with the silicon-based CMOS industry. The robust tunneling numerical methods for simulation need to be worked on so as to minimize the device development cycle time and catch up with the virtual fabrications available for CMOS. The revolution in providing high-end non-CMOS materials can play a vital role as the presence of traps can impact the electrostatic control and device performance.

1.7 SUMMARY

TFETs show tremendous potential from a low-power application point of view. However, the challenge of gaining a high Ion/Ioff ratio in combination with a subthreshold swing <60mV/decade still persists for devices to be used in high-performance circuits. To achieve this, it is required that many technologies be used in combination to realize such heterostructure TFET devices. In lower node technology, its fabrication becomes viable, and TFETs can be real, standalone devices comparable to MOSFETs.

BIBLIOGRAPHY

Avci, Uygar E., Daniel H. Morris, and Ian A. Young, "Tunnel Field-effect Transistors: Prospects and Challenges", IEEE Journal of the Electron Devices Society, Vol. 3, No. 3, PP. 88–95, 2015.

Bhardwaj, A.K., S. Gupta, B. Raj, and Amandeep Singh, "Impact of Double Gate Geometry on the Performance of Carbon Nanotube Field Effect Transistor Structures for Low Power Digital Design", Computational and Theoretical Nanoscience, ASP, Vol. 16, PP. 1813–1820, 2019.

Boucart, Kathy, and Adrian Mihai Ionescu, "Double-gate Tunnel FET with High-k Gate Dielectric", IEEE Transactions on Electron Devices, Vol. 54, No. 7, PP. 1725–1733, 2007.

Chawla, T., M. Khosla, and B. Raj, "Optimization of Double-gate Dual Material GeOI-Vertical TFET for VLSI Circuit Design", IEEE VLSI Circuits and Systems Letter, Vol. 6, No. 2, PP. 13–25, August 2020.

Datta, Suman, Huichu Liu, and Vijaykrishnan Narayanan, "Tunnel FET Technology: A Reliability Perspective", Microelectronics Reliability, Vol. 54, PP. 861–874, 2014.

Dutta, Ritam, T.D. Subash, and Nitai Paitya, "Improved DC Performance Analysis of a Novel Asymmetric Extended Source Tunnel FET for Fast Switching Application", Silicon, Springer Nature B.V., 2021.

Goyal, C., J. Subhi, and B. Raj, "Low Leakage Zero Ground Noise Nanoscale Full Adder using Source Biasing Technique", Journal of Nanoelectronics and Optoelectronics, American Scientific Publishers, Vol. 14, PP. 360–370, March 2019.

Goyal, C., J.S. Ubhi, and B. Raj, "A Low Leakage CNTFET Based Inexact Full Adder for Low Power Image Processing Applications", International Journal of Circuit Theory and Applications, Wiley, Vol. 47, No. 9, PP. 1446–1458, September 2019.

Goyal, C., J.S. Ubhi, and B. Raj, "A Reliable Leakage Reduction Technique for Approximate Full Adder with Reduced Ground Bounce Noise", Journal of Mathematical Problems in Engineering, Hindawi, Vol. 2018, PP. 16, Article ID 3501041, 15 October 2018.

Ionescu, Adrian M., and Heike Riel, "Tunnel Field-effect Transistors as Energy Efficient Electronic Switches", Nature, Vol. 479, No. 7373, P. 329, 2011.

Jain, A., S. Sharma, and B. Raj, "Analysis of Triple Metal Surrounding Gate (TM-SG) III-V Nanowire MOSFET for Photosensing Application", Opto-Electronics Journal, Elsevier, Vol. 26, No. 2, PP. 141–148, May 2018.

Jain, N., and B. Raj, "Parasitic Capacitance and Resistance Model Development and Optimization of Raised Source/Drain SOI FinFET Structure for Analog Circuit Applications", Journal of Nanoelectronics and Optoelectronins, ASP, USA, Vol. 13, PP. 531–539, April 2018.

Kaur, M., N. Gupta, S. Kumar, B. Raj, and Arun Kumar Singh, "RF Performance Analysis of Intercalated Graphene Nanoribbon Based Global Level Interconnects", Journal of Computational Electronics, Springer, Vol. 19, PP. 1002–1013, June 2020.

Kim, Sung Hwan, Germanium-Source Tunnel Field Effect Transistors for Ultra-low Power Digital Logic. University of California, 2012.

Kumar, Satyendra, Kaushal Nigam, Saurabh Chaturvedi, Areeb Inshad Khan, and Ashika Jain, "Performance Improvement of Double-Gate TFET Using Metal Strip Technique", Silicon, Vol. 14, PP. 1759–1766, 2021. Springer Nature B.V.

Kumar, V., S. Kumar, and B. Raj, "Design and Performance Analysis of ASIC for IoT Applications", Sensor Letter, ASP, Vol. 18, PP. 31–38, January 2020.

Reddy, N. Nagendra, and Deepak Kumar Panda, "A Comprehensive Review on Tunnel Field-effect Transistor (TFET) Based Biosensors: Recent Advances and Future Prospects on Device Structure and Sensitivity", Silicon, Springer Nature, B.V., Vol. 13, PP. 3085–3100, 2020.

Saeidi, Ali, M. Ionescu, et al., "Nanowire Tunnel FET with Simultaneously Reduced Subthermionic Subthreshold Swing and Off-current Due to Negative Capacitance and Voltage Pinning Effects", Nano Letters, Vol. 20, PP. 3255–3262, 2020.

Saurabh, Sneh, and Mamidala Jagadesh Kumar, Fundamentals of Tunnel Field effect Transistors. CRC Press, 2016.

Sharma, S.K., B. Raj, and M. Khosla, "Enhanced Photosensivity of Highly Spectrum Selective Cylindrical Gate In1-xGaxAs Nanowire MOSFET Photodetector", Modern Physics Letter-B, Vol. 33, No. 12, PP. 1950144, 2019.

Singh, A., M. Khosla, and B. Raj, "Design and Analysis of Dynamically Configurable Electrostatic Doped Carbon Nanotube Tunnel FET", Microelectronics Journal, Elesvier, Vol. 85, PP. 17–24, March 2019.

Singh, G., R.K. Sarin, and B. Raj, "Fault-Tolerant Design and Analysis of Quantum-Dot Cellular Automata Based Circuits", IEEE/IET Circuits, Devices & Systems, Vol. 12, PP. 638–664, 2018.

Singh, J., and B. Raj, "Design and Investigation of 7T2M NVSARM with Enhanced Stability and Temperature Impact on Store/Restore Energy", IEEE Transactions on Very Large Scale Integration Systems, Vol. 27, Issure 6, PP. 1322–1328 June 2019.

Singh, J., and B. Raj, "Modeling of Mean Barrier Height Levying Various Image Forces of Metal Insulator Metal Structure to Enhance the Performance of Conductive Filament Based Memristor Model", IEEE Nanotechnology, Vol. 17, No. 2, PP. 268–267, March 2018 (SCI).

Singh, S., S. Bala, B. Raj, and B. Raj, "Improved Sensitivity of Dielectric Modulated Junctionless Transistor for Nanoscale Biosensor Design", Sensor Letter, ASP, Vol. 18, PP. 328–333, April 2020.

Singh, S., and B. Raj, "A 2-D Analytical Surface Potential and Drain Current Modeling of Double-Gate Vertical T-shaped Tunnel FET", Journal of Computational Electronics, Springer, Vol. 19, PP. 1154–1163, April 2020.

Singh, S., and B. Raj, "Analytical Modeling and Simulation Analysis of T-shaped III-V Heterojunction Vertical T-FET", Superlattices and Microstructures, Elsevier, Vol. 147, PP. 106717, November 2020.

Singh, S., and B. Raj, "Design and Analysis of Hetrojunction Vertical T-shaped Tunnel Field Effect Transistor", Journal of Electronics Material, Springer, Vol. 48, No. 10, PP. 6253–6260, October 2019.

Singh, S., and B. Raj, "Modeling and Simulation Analysis of SiGe Hetrojunction Double Gate Vertical T-shaped Tunnel FET", Superlattices and Microstructures, Elsevier, Vol. 142, PP. 106496, June 2020.

Tajally, M.B., and M.A. Karami, "TFET Performance Optimization Using Gate Work Function Engineering", Indian Journal of Physics, Vol. 93, No. 9, PP. 1123–1128, September 2019.

Wadhera, T., D. Kakkar, G. Wadhwa, and B. Raj, "Recent Advances and Progress in Development of the Field Effect Transistor Biosensor: A Review", Journal of Electronic Materials, Springer, Vol. 48, No. 12, PP. 7635–7646, December 2019.

Wadhwa, G., and B. Raj, "An Analytical Modeling of Charge Plasma based Tunnel Field Effect Transistor with Impacts of Gate Underlap Region", Superlattices and Microstructures, Elsevier, Vol. 142, PP. 106512, June 2020.

Wadhwa, G., and B. Raj, "Design and Performance Analysis of Junctionless TFET Biosensor for High Sensitivity", IEEE Nanotechnology, Vol. 18, PP. 567–574, 2019.

Wadhwa, G., and B. Raj, "Label Free Detection of Biomolecules Using Charge-Plasma-Based Gate Underlap Dielectric Modulated Junctionless TFET", Journal of Electronic Materials (JEMS), Springer, Vol. 47, No. 8, PP. 4683–4693, August 2018.

Wadhwa, G., and B. Raj, "Parametric Variation Analysis of Charge-Plasma-based Dielectric Modulated JLTFET for Biosensor Application", IEEE Sensor Journal, Vol. 18, No. 15, 1 August 2018.

Yadav, D., S.S. Chouhan, S.K. Vishvakarma, and B. Raj, "Application Specific Microcontroller Design for IoT Based WSN", Sensor Letter, ASP, Vol. 16, PP. 374–385, May 2018.

Zhang, M., et al., "Simulation Study of the Double-gate Tunnel Field-effect Transistor with Step Channel Thickness", Nanoscale Research Letters, Vol. 15, P. 128, 2020.

Chapter 2

Fundamentals of TFETs and Their Applications

V. Ramakrishna and A. Krishna Kumar

CONTENTS

2.1 INTRODUCTION

The metal–oxide–semiconductor field-effect transistor (MOSFET) has played a significant role in semiconductor engineering during the last few decades [1]. However, the size of components has been scaled down according to Moore's law, and many devices have been encountered. Difficulties

DOI: 10.1201/9781003311379-2

need to be resolved. The tunneling field-effect transistor (TFET) is seen as one component that has significant potential for MOSFET replacement, as the TFET has a subthreshold swing (SS) of less than 60 mV/dec threshold and a very low leakage [2]. At present, these characteristics are conducive to the miniaturization of Drain Voltage (VD) and reduce the problem of energy consumption, so it is suitable for application in low-power components.

However, the current research literature points out that the biggest fatal flaw of TFETs is that the current value in the on-state is too low, which greatly limits the development of TFETs. Therefore, this study uses a semiconductor simulation tool (technology computer-aided design, TCAD) [3] to design a structure with a vertical tunneling mechanism and discusses the optimization of component parameters and the use of heterogeneous junctions in n-type and p-type TFETs to improve TFET performance and finally integrate the structure with vertical tunneling and lateral tunneling mechanisms. The results show that having a large-area vertical tunneling mechanism can indeed effectively increase the tunneling current. The heterogeneous junction formed by silicon germanium and silicon can also help the tunneling effect. Finally, the two tunneling mechanisms are integrated on the same device. The result also shows that the overall on-current of the component can be increased, which helps overcome the problem of the too-small on-current value of the TFET [4].

2.1.1 Research Background and Motivation

In the era of rapid technological change, a variety of electronic products are moving towards the development of light, thin, short, and multifunctional. Integrated circuits (ICs) have continued to be developed and innovated in the past fifty years, following Moore's law. The direction of Moore's law keeps shrinking the size of the transistor so that the transistor can continue to increase its density, improve performance, and increase applications. But today, the component scaling of complementary metal–oxide semiconductors (CMOSs) has been getting closer and closer to the limit of the physical scale of semiconductors, and many difficulties have been encountered in the manufacturing process. In addition to scaling down, many problems that have not been encountered in the past were also found in the research process. In addition to the short channel effects (SCEs), gate oxide leakage current, and the improvement of efficiency, the area is reduced. The threshold voltage (V_{th}) limit of the component will make the component more difficult to control, and the resulting electrical disturbance of the component is also widely noticed. Therefore, how to overcome or find alternative solutions is also one of the important issues of concern at present.

2.1.2 Research Direction

The driving force behind the development of integrated electronic technology has been its search for high-performance computers, improved networking devices, and consumer products such as cell phones. Transistor scaling has been the key development in electronics in addition to raising the chip size and an improved logic architecture to maximize efficiency, speed, and lower energy and expenses. Since the size of the feature is reduced, more transistors can be combined into a chip, which can, in turn, reduce costs and improve performance.

The current depends on thermionic emission in conventional MOSFETs from the carrier across a potential barrier. As such, it is one of the greatest challenges for traditional materials at advanced technical nodes to sustain their power usage within a reasonable range. TFETs have various intriguing electrical features due to their diverse current transmission mechanisms and may be a workable alternative to traditional MOSFETs. This section discusses the TFETs' configuration and operational theory, as well as the origin and existence of some of TFETs' most distinguishing electrical properties. The tunnel current, as well as the physical and electrical characteristics of TFETs, are closely tied. This enables readers to comprehend how system parameters may be used to configure TFETs. Finally, this chapter highlights the benefits and drawbacks of TFETs, as well as briefly discussing why TFETs could be a viable alternative to regular MOSFETs [5]. As nanometer scales are met, power usage becomes a key bottleneck for further scaling. The continuously decreasing MOSFET size causes increased leakage current due to drain-induced barrier-lowering SCEs, which refers to the transistor threshold reduction at higher drain voltages, and the power supply voltage cannot be lowered anymore because the subthreshold slope is only 60mV/decade at room temperature [6]. In this regard, it's crucial to look at alternative devices that might outperform MOSFETs with these nanometer measurements.

The tunneling effect is a common phenomenon when transistors are shrunk to the nanoscale size. Most of the negative effects that accompany transistors are mostly negative effects. A TFET is the application of the principle of tunneling as a mechanism for controlling the switching of transistors and has been regarded as a very promising element in recent years. The reason is that a TFET is different from the operating mechanism of MOSFET, so it can avoid many short channel and reliability problems encountered when shrinking the size. A TFET has an SS of less than 60mV/dec and a very small leakage current. These features are conducive to the miniaturization of VD and reduce the problem of energy consumption, so it is suitable for application in low-power components. However, TFET also has some problems that need to be overcome, such as low on-state current value, high concentration of doping, and high accuracy of doping position. These are the goals that need to be solved when developing TFETs in the future.

2.2 POWER ISSUES OF MOSFET

Microprocessor power, density, networking, and costs have all increased significantly as the size of CMOS field-effect transistors has shrunk. Advanced CMOS technologies, on the other hand, are now confronted with two problems, all of which result in high power consumption: the difficulties in further decreasing the power supply and stopping rising flows that reduce the switching ratio of "on" and "off" (ION/IOFF) currents [7].

A MOSFET switching process involves electron injection thermionic (temperature-dependent) through an energy barrier. The steepness of the transformation path from start to finish sets a simple upper limit. The voltage of the gate required to adjust the drain current by an order of quantity is represented in the expression SS, S: when the transistor operates in the subthreshold sector. Reducing the size of transistors has been one of the most important development strategies of semiconductor technology since its beginning. More transistors can be placed onto a smaller chip as the transistor size is reduced, so the chip does not need to be large for practical enhancement.

Furthermore, it will improve the processor's computing performance while lowering the chip's power consumption, allowing it to satisfy the demands of upcoming mobile devices. When the MOSFET size is repeatedly reduced, the transistor can run into quantum physics problems, allowing the transistor to leak power, negating the original gain of decreasing the channel duration. Various methods for improving the bottleneck encountered in the reduction of transistor size, high performance, and density have been established in the last ten years to improve this challenge, as shown in Figure 2.1 [8].

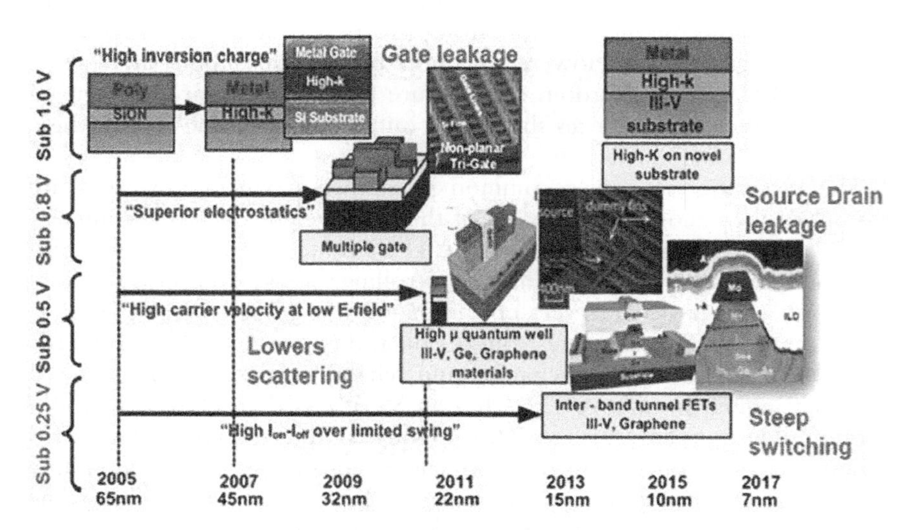

Figure 2.1 Development direction of high-performance transistor innovation. Transformative changes in the transistor architecture (3D, TFET) [8].

One method is to use the strain techniques to effectively increase the drive current of CMOS, and to increase the carrier mobility of the MOSFET. Strain techniques, like SiGe incorporation, should enhance performance in forthcoming generations of CMOS transistors without deeply scaling transistor dimensions. The second method is to use a high dielectric constant (high-k) material as the gate oxide layer to replace the traditional silicon dioxide (SiO_2) material, which can effectively resist the deposition of the same equivalent oxide layer thickness. The leakage current from the gate terminal tunneling successfully solves the problem of gate leakage current and further introduces the metal gate process to replace the traditional poly-gate process. Because the metal gate process can effectively reduce the resistance of the gate terminal electrode, the factors of Resistive-Capacitive (RC) delay can be reduced. The third method is to develop three-dimensional (3D) structure transistors to replace traditional planar transistors.

The fourth method is to use materials with high electron mobility to form the transmission channel of the device, such as strained silicon (strain material, germanium material, and III–V group material), can reduce the operating voltage of the device itself so that the electric field will not be too large. It causes the transistor to collapse and has a chance to get a larger on-current. The preceding method can reduce the operating voltage of the device and reduce the power loss and leakage current of the transistor. However, the previously mentioned methods still cannot effectively solve the problem of transistor size reduction, mainly because the MOSFET uses drift diffusion to dominate the switching current characteristics of the device as shown in Figure 2.2. Under ideal conditions at room temperature, the SS can only drop to a minimum of 60 mV/decade. Taking into account the leakage current IOFF and power loss in the off state, the operating voltage (<0.5 V) [10] of the MOSFET will quickly exceed the power loss range that the circuit can tolerate.

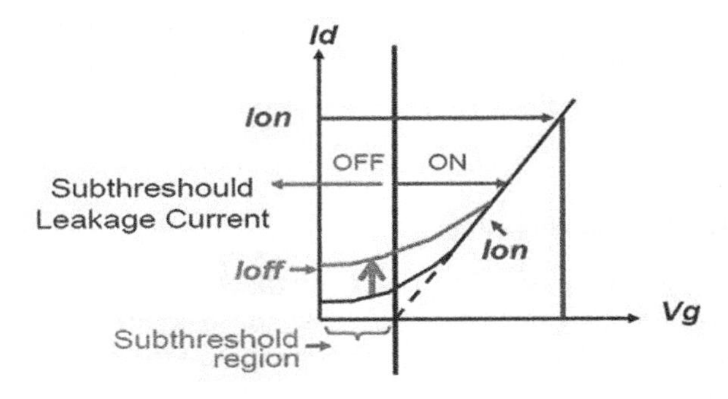

Figure 2.2 Subthreshold leakage current of a MOSFET [9].

TFETs will have the opportunity to become an important direction for the development of transistor size reduction in the future. Compared with MOSFETs, they mainly have the following advantages such as lower secondary threshold swing (<60 mV/decade), better switching characteristics, smaller operating voltage, lower power loss, and small leakage current. The TFET itself uses the tunneling of carriers between energy bands to generate conduction current, so the secondary limit swing of the element will not be limited to 60 mV/decade at room temperature due to the influence of temperature, it can be low. The low operating voltage also has a smaller shutdown current, providing ultra-low power loss to achieve good energy efficiency. In order to break through the problem of low on-current, III–V group materials are used to make TFETs. Compared with group IV materials, group III–V materials have an effective tunneling barrier with a smaller band adjustment space, so tunneling current can be obtained more efficiently, although the on-current (Ion) increases, also because the tunneling energy barrier becomes smaller, in the off state. In the off state, the leakage current of the component (Ioff) also increases, so how to make the component have a high on-current while suppressing the leakage current in the off state is one of the important challenges faced by the current use of III–V materials for TFETs.

2.2.1 Basics of TFETs

Today, the size of the transistor has been shrinking, and it has gradually approached the limit of physical size, and the manufacturing process is also encountered many difficulties that need to be resolved, so in recent years, many research teams have begun to seek other alternatives. Among them, TFETs are regarded as a kind with a potential component. The biggest feature is that TFETs have a SS of less than 60 mV/dec and a very small leakage current; these features are conducive to the miniaturization of VD and reduce energy consumption [11–13].

Therefore, it is suitable for low-power components, as shown in Figure 3.3. However, TFET also has some problems that need to be overcome, such as low on-state current value, high concentration of doping, and high accuracy of doping position. These are the goals that need to be solved when developing TFETs in the future.

2.2.2 Basic Characteristics

Figure 2.4 shows the basic n-type and p-type TFET structures. From the figure, note that there is not much difference in structure between a general MOSFET and a TFET. The main dissimilarity lies in the type of doping. The drain and source doping types are similar, but in TFETs, the source and drain doping types are reversed. In addition, low-concentration doping is used in

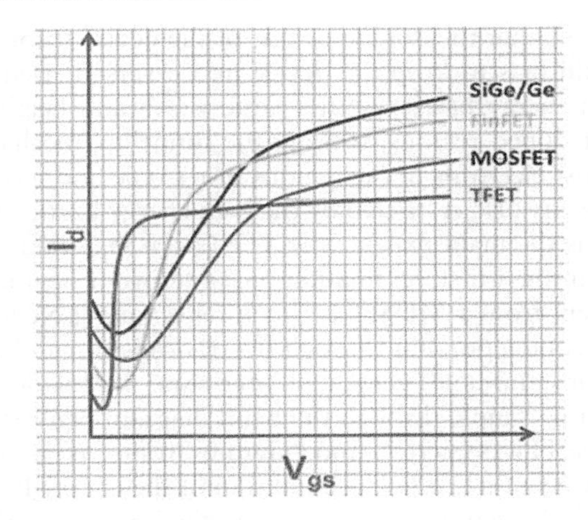

Figure 2.3 TFET and its transfer characteristics.

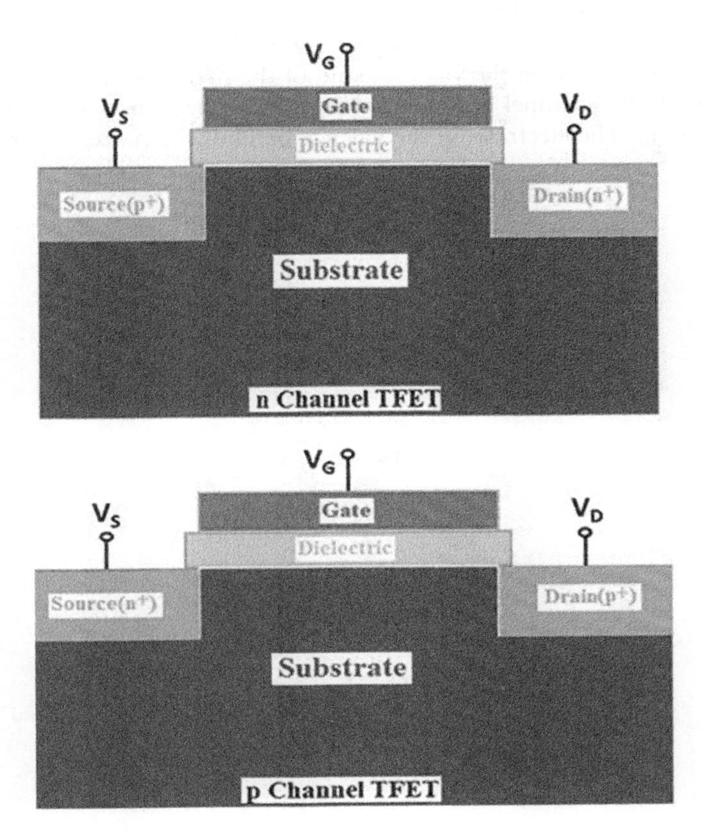

Figure 2.4 n-type and p-type TFET structures.

the channel area so that the doping profile of the device forms a p-i-n form, with proper drain bias to keep the device in a reverse-biased state. At this time, there is an energy barrier with a large width and height at the junction of different doped regions, making the tunneling mechanism difficult to occur, and most electrons in the regions of n+-type and most holes in the region of p+-type are also restricted by the high energy barrier and cannot be turned on. This is the main reason why TFETs can maintain extremely low leakage current in the off state. In addition, in terms of operation principle, a TFET is different from a MOSFET, which relies on a large number of carriers to conduct conduction but through the transmission and conduction of tunneling electrons and holes generated after band modulation. The tunneling principle can be simple through the band diagram Figure 2.5 shows the n-type and p-type TFETs energy band diagrams, respectively [14, 15].

Consider the case of nTFET. The condition of the energy band before the gate is biased is indicated by the solid line in the image, and the TFET is in the off state at this point. The energy band of the channel area is impacted by the bias voltage and lowers when a sufficiently strong positive bias is given to the gate, as indicated in the picture by the red dotted line.

The distance between the valence band of the p+ region and the conductive band of the channel is shortened at the channel junction surface and the p+ region. The electrons in the valence band of the p+ region cross the

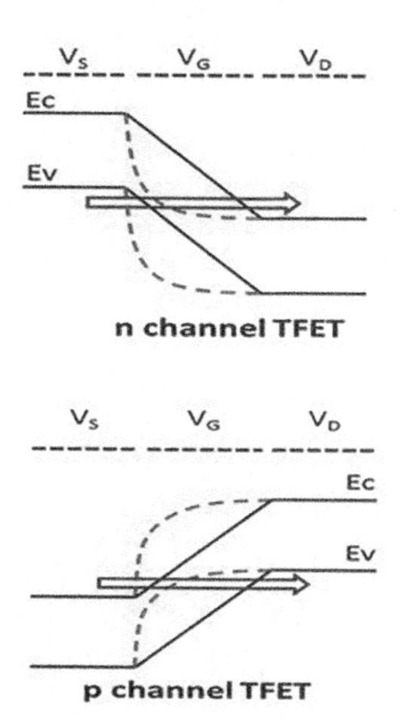

Figure 2.5 n-type and p-type TFET energy band distribution diagram.

energy gap into the channel's conductive band and flow into the n+ region. The preceding behavior is called the tunneling effect. In a general MOSFET, the source area of the main carrier as defined as the source, and the area receiving the carriers is defined as the drain. TFETs also use a similar concept. The p+ doped region is defined as the source in nTFETs.

The drain is defined as the n+ doped region. In contrast, if a sufficiently enough negative bias voltage is supplied to the gate in a pTFET, the bias voltage will impact the channel area energy band and cause it to increase, as indicated in the top dotted line in the figure, at this moment in the channel. The distance between the channel valence band and the n+ conductive band region is shortened near the surface and the n+ region, allowing valence band electrons to pass through the energy gap and reach the conductive band of the n+ region, while holes flow in the p+ region; thus, the n+ doped region is defined as the source, and the p+ doped region is defined as the drain in a pTFET [16–19].

2.2.3 Basic Tunneling Theory

The conduction condition of the TFET is defined as the timing of the tunneling behavior. As a result, the tunneling probability of electrons is strongly connected to the device's output properties. From the reference, it can be known that the tunneling probability of electrons in the on state between energy bands is as follows as shown by Patel et al. [20]:

$$T(E) = \exp\left(-\frac{4\sqrt{2m^*}E_g^{\frac{3}{2}}}{3|e|\bar{h}\left(E_g + \Delta\varnothing\right)}\sqrt{\frac{\varepsilon_{si}}{\varepsilon_{ox}}t_{ox}t_{st}}\right)\nabla\varnothing\,. \quad (2.1)$$

Among these, m^* represents the carrier effective mass, E_g represents the energy gap (band gap), e represents the electron charge, and h represents the reduced Plank's constant. The dielectric constants and thicknesses of the silicon and oxide layers, respectively, are Si, ox, t_{Si}, and t_{ox} for the relative energy-gap difference where tunneling occurs [21, 22].

It can be observed from the preceding parameters that reducing the equivalent quality, using high-permittivity materials, and reducing the thickness of the oxide layer are the same as the optimization guidelines of traditional MOSFETs (improving carrier mobility and increasing oxide layer capacitance), with energy gap. Small materials can also effectively increase the tunneling probability. Furthermore, the gate bias voltage controls the relative energy gap difference at the region where tunneling occurs, and the gate bias has an exponential influence on the tunneling probability. This means that when the device goes from the off state to the on state, the tunneling probability will increase exponentially. As a result, the tunneling current will also be generated in a large amount instantaneously, so the TFET can have a relatively small SS.

Also because the working principle of a TFET and a traditional MOSFET is not the same, the common saturation current formula is not applicable to a TFET, but the tunneling current is used to express it, as shown in Equation 2.2 [23]:

$$I_t = a.AV_{eff}\varepsilon.e^{\left(-\frac{B}{\varepsilon}\right)} \tag{2.2}$$

$$A = \frac{\sqrt{2m^*}q^3}{4\pi^2\bar{h}^2E_g^{\ 1/2}} \tag{2.3}$$

$$B = \frac{4\sqrt{2m^*}E_g^{\ 3/2}}{3q\bar{h}} \tag{2.4}$$

Among them, V_{eff} is the relative Fermi-level difference where tunneling occurs, ε is the electric field, and A and B are the former and exponential terms, respectively, and then the formula (Equation 2.2) is simplified into the Taylor expansion to get

$$I_t = a.AV_{eff}\varepsilon(1 - \left(\frac{B}{\varepsilon}\right) = a.AV_{eff}(\varepsilon - B). \tag{2.5}$$

It can be seen from the preceding formula that the tunneling current is related to the tunneling area. If the area factor is not considered first, Equation 2.5 is analyzed and sorted out after unit analysis.

$$I_t = amp.cm^{-2} = C_{ox}V_{sat}(V_g - V_T)\frac{1}{cm} \tag{2.6}$$

It can be seen from the preceding formula that the tunneling current is related to the tunneling area. From the preceding formula, it can sort out that when the tunneling area and component width are not considered, the similarity between the tunneling current and the MOSFET current can be obtained [24–27].

The main goal of current TFET research is how to effectively increase the on-current without losing the original advantages of TFET (low leakage current and small SS). There are two main research directions, one of which is to use two different materials in the tunnel area to form a heterojunction at the junction of the material so that the energy band distribution at the junction can be relatively large. The height difference can reduce the height and width of the tunneling barrier of the device in the on state, increase the possibility of the tunneling mechanism, and effectively increase the on-state current, but on the contrary, it will also be the same in the off state. The reason causes the leakage current to increase slightly. Common combinations, such as the combination of silicon germanium and silicon carbide; the other is to improve the design of the device structure to increase the area of the tunneling area. This method can also improve the open state. Current, due to the principle of device miniaturization, is relatively impractical when increasing the size of the device in order to increase the tunneling area. Therefore, it is hoped that this goal can be achieved through different design structures.

2.3 TYPES OF TFETS

TFETs can be categorized into two groups depending on their structure: planar and 3D structures [28]. A planar TFET is one that has a planar current-carrying surface. A silicon-on-insulator (SOI) wafer or a bulk silicon wafer may be used to build the unit. SOI TFETs are favored over bulk TFETs for improved gate control over the tube, and only the former has been extensively tested. Let's look at some of the most significant planar TFET structures first [29, 30].

2.3.1 Double-Gate TFETs

The fundamental design is a gated p-i-n diode. It has two gates: one above (the front gate) and the other below (the back door). It has two gates (called the back gate). By terminating field lines from the gate, this procedure enhances electrostatic safety at the door of the channel. The ON-state stream is enhanced over a single TFET gate since the device has two channels through which current can flow. The scheme of p-channel double-gate (DG)-FET [31] is shown in Figure 2.6.

Because two gate voltages are used in a DG-FET, the current is doubled. As a result, the ON-current is greater than the OFF-current of a single gate TFET. For varying levels of V_{ds} voltage and I_{off}, the TFET decreases. When compared to MOSFETs, this is a significant achievement in terms of energy savings in low-power devices.

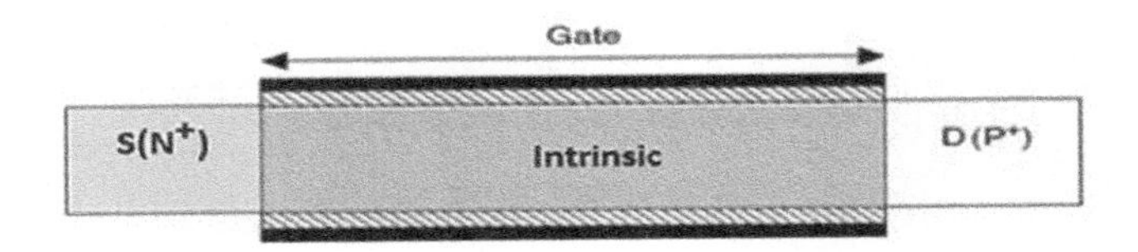

Figure 2.6 p-channel double-gate schematic.

2.3.2 Planar TFETs

A planar TFET is a device having a planar current-carrying surface. This device may be built on a bulk silicon wafer or an SOI wafer. SOI TFETs are preferred over bulk TFETs for improved gate control over the channel. The surface tunnel transistor was the first tunnel transistor to cope with the speed, power, and the I_{OFF}/I_{ON} ratio. The first p-i-n basic TFET structure is what controls the speed, power, I_{OFF}/I_{ON}, tuning range, and so on. A thin film of silicon (usually approximately 10 nm or less) is formed on a layer of buried oxide in the form of an SOI TFET [32] on a silicon substrate (roughly 100 nm thick). The gate oxide is grown (1–2 nm thick). Figure 2.7 shows the p-channel p-i-n TFET schematic diagram.

2.3.3 Dual TFETs

The dual-material-gate (DMG) TFET [33] configuration is a significant modification of the TFET structure. It is composed of two gates with distinct work functions that run the length of the channel, one covering the area near the drain and the other covering the area near the source.

Tunneling occurs at the source–channel junction while the system is in ON mode; thus, the source gate is referred to as the tunneling gate. The auxiliary gate is the gate closest to the drain. Sidewall spacer methods can be used to build the DMG frame. Figure 2.8 depicts a cross section of a DMG

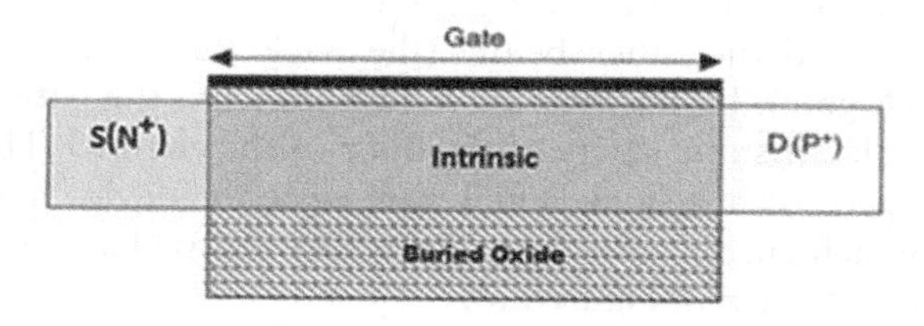

Figure 2.7 p-channel p-i-n TFET schematic.

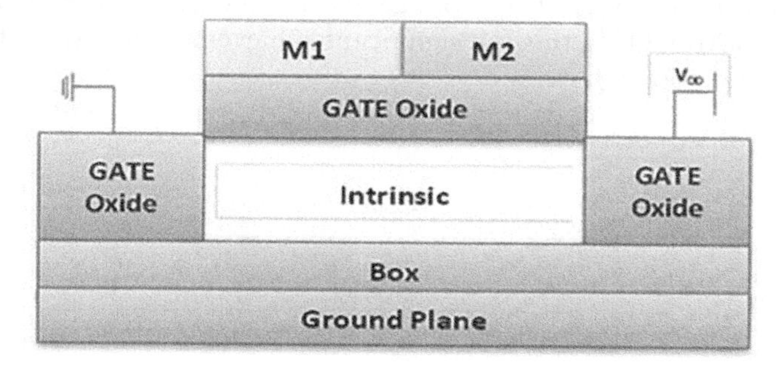

Figure 2.8 DMG TFET cross section.

TFET. The source and drain are made up of highly doped p-type and n-type regions, respectively. A substantially doped n-type layer makes up the intermediate channel area. The gate dielectric is silicon dioxide (SiO_2). The gate is made up of two materials, M1 and M2, and has L1 and L2 gate lengths as well as m1 and m2 work functions. The device is classified as an n-type TFET or a p-type TFET depending on whether a positive or a negative voltage is applied to the gate terminal. The transistor functions as an n-TFET when a positive gate voltage is supplied and as a p-TFET when a negative gate voltage is applied [34].

2.3.4 Raised Germanium-Source TFETs [35]

The usage of germanium (Ge) in sources is widely distributed due to the low ON current of silicone-based TFETs, due to the narrow band gaps of Ge. Given the fact that tunneling is the characteristic feature of the electric field, it's also a possible option to change the TFET structure in line with the carrier tunnel direction. As a result, higher TFET Ge sources are key to TFET structures, with several benefits over a standard TFET structure based on Si. The high source system offers a broader tunnel area than the Si TFET standard, whereby tunneling occurs only in a limited area close to the source [36].

2.3.5 Heterojunction TFET

A heterojunction is a type of semiconductor material that has uneven band gaps between two layers of different semiconductors. Heterojunctions operate at high frequencies and are also found in high-electron-mobility transistors (HEMTs). When the band gaps of crystalline semiconductors are uneven, TFETs function at low voltage. There are three forms of heterojunction semiconductor alignment, and they are straddling gaps, staggered gaps, and broken gaps. The heterojunction TFET [37], which is made up of III–V components, is another TFET system for achieving a higher ON-current.

2.3.5.1 Optimization of TFETs Using Heterojunction

General TFET components are made of the same material, so the region where tunneling occurs belongs to the homojunction. The energy gaps on both sides of the junction are equal, and the relative position is adjusted by different doping types and concentrations. A heterojunction is a junction formed by two materials with different energy gap sizes and similar lattice sizes. Figure 2.9 shows the three common alignment methods for heterojunctions. The heterojunctions on the left and right are not suitable for use in TFETs, because the energy band distribution on the left requires a very large gate bias to occur at the junction. In the tunneling mechanism,

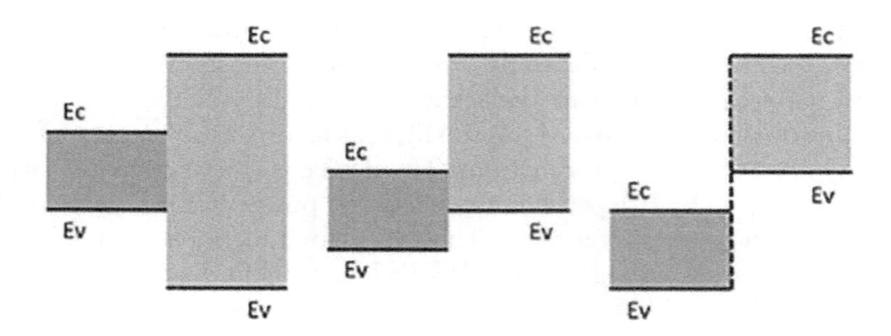

Figure 2.9 Schematic diagram of different forms of heterogeneous junction.

on the contrary, the gap of the right junction energy gap is too large, which will cause the TFET to turn on without the gate bias, so it is not a suitable shape; only the middle energy band distribution is an ideal state. Since the tunneling probability has a considerable relationship with the position of the energy band distribution, if a suitable material is selected and doped to form a heterogeneous junction, the performance of the TFET can be greatly improved [38, 39].

If we consider homojunction, TFETInAs is a homojunction TFET device since it contains InAs material in all its source, gate, and drain regions. The InAs TFET device is built using the same parameters as the Si-based TFET device. Lower band-gap materials can be utilized instead of Si-based TFET devices to increase the device's ON current. Lower band-gap materials, such as InAs, are utilized for the source, the drain, and the channel, which promotes tunneling and therefore improves the ON current and SS of the TFET device [40–43].

2.4 TFETS' APPLICATIONS

TFETs for many applications, including conventional logical gates, memory, sensors, and analog circuits, are being explored. In various new experiments, TFET-based circuits have been proved to outperform standard low-voltage CMOS circuits, in particular regarding energy efficiency. These discoveries led researchers to study circuits that can preserve energy and power using the particular characteristics of TFETs. New TFET-based circuit implements are conceivable as they have considerably different electrical characteristics than a conventional MOSFET.

TFETs have other electrical features, including unidirectional current flow and exponential drain current input, not found in regular MOSFETs. In TFET circuit applications, these electrical characteristics offer additional challenges. It is not possible simply to make a CMOS circuit functional by substituting MOSFETs with equivalent TFETs. Researchers are thus examining the challenges of TFET use in circuits by looking at novel circuits and TFET designs. The following sections give various TFET applications for digital circuits, memory, and analog circuits. TFETs were initially investigated as a replacement for conventional MOSFETs in digital applications and logical applications. However, the advantages of using TFETs in a variety of analog circuits have recently been investigated. Because of their minimal SS and excellent saturation in the output characteristics, TFETs have been demonstrated in these experiments to be attractive candidates for low-power analog circuits. However, the exponential beginning of output characteristics, unidirectional current flow, and delayed saturation of output characteristics found in TFETs might have a negative impact on some analog circuits. The special electrical characteristics of TFETs, on the other hand, can be used in analog circuits to create novel topologies. TFETs' inherent characteristics are important to analog systems, such as transconductance, transconductance efficiency, output strength, and intrinsic voltage gains. TFETs are comparable to regular MOSFETs in amplifier circuits. Si TFETs have recently been compared to a single-circuit CMOS implementation of common source amplifiers. The SiTFET common source amplifier's high output resistance has been found to result in an even larger power gain when compared to a comparable CMOS amplifier. Because of the smaller gain (gm) of Si TFET and its higher total capacity, the Si TFET common source amplifier has a narrower bandwidth than a comparable CMOS amplifier. The high output resistance of the TFET can also be exploited to simplify TFET installation by eliminating cascode transistors with a cascode current mirror. The removal of the cascode transistors reduces the power supply and relaxes the minimum voltage constraint on the power output voltage. A gate stack design may be utilized to study analog performance, while the performance of TFET devices is mainly focused on digital applications. TFETs are becoming a viable option for analog applications.

2.4.1 TFET Applications in Digital Circuits

Figure 2.10 depicts the circuit diagram of a TFET-based inverter. It consists of a pull-down network (PDN) formed by an NTFET and a pull-up network (PUN) formed by a PTFET (PUN).

A PTFET's drive strength is generally smaller than that of an NTFET. As a result, an inverter with a lower driving power PTFET exhibits symmetrical voltage transfer characteristics (VTC). To enhance PTFET current or make VDD/2 transmission characteristics symmetrical and transform several PTFETs on the PUN, device-level methods are used. TFETs are examined for usage in digital circuits other than inverters. The TFETs may be used to apply pass transistor logic (PTL). However, there are numerous issues with PTL standard implementation when utilizing TFETs owing to its unidirectional current flow in TFETs. These issues may be illustrated as composed of

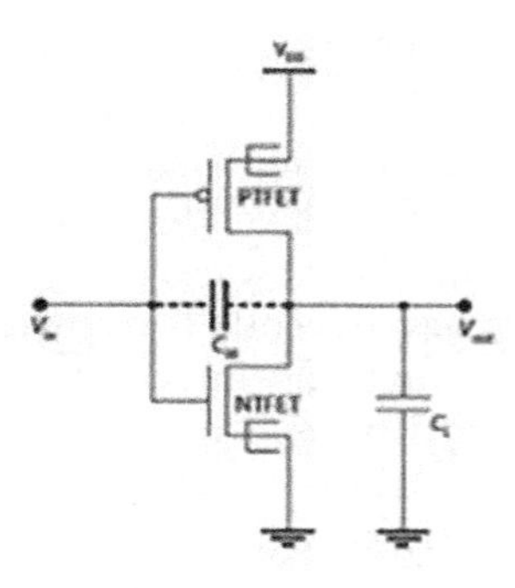

Figure 2.10 TFET inverter circuit diagram.

an NTFET and the PTFET, where the input and output drains of transistors are linked. The gate of an NTFET is connected to the choose signal, whereas the gate of a PTFET is connected to the gate signal.

Depending on the polarity of the drain to source voltages, only one TFET (both NTFET and PTFET) can operate at high-signal pickups. If the VDS is positive, the NTFET is positive, and the PTFET is negative. In comparison, both PMOS and NMOS lead when the signal is selected to be high and lead to a high drive current as well as to a solid "0" and "1" logic when using standard MOSFET.

2.4.2 TFET Applications in Memories

Inter-band tunneling transistors have shown that they can be used to replace MOSFETs in low-power memory circuits. The fundamental property of these transistors is that they have a very low OFF current. Low OFF current levels in memory cells in standby mode can result in substantial power savings. Because memory consumes the bulk of the processor's on-chip area, power reductions in this module can help improve overall system performance. Due to the exceptionally low leakage current and good subthreshold swings in Static Random Access Memory (SRAMs), TFETs are intensively investigated for use [44]. However, unidirectional TFET current conduction and delayed saturation in output properties present issues when implementing SRAM circuits based on TFETs. The circuit has the architecture of a typical CMOS-SRAM 6-transistor (6T) technology and the integration of TFETs in the SRAM system is a challenge. If there is no power-saving action on the CPU, the SRAM cell falls into standby mode. The device used to generate SRAM requires lower power in sleep mode in order to optimize the overall power consumption of the memory cell. Tunneling devices have an extremely low OFF current, offering them a competitive edge compared to standard MOSFETs.

2.4.3 Other Applications of TFETs

TFETs have mostly been explored for digital applications and in logical applications as a replacement for conventional MOSFETs. Nevertheless, the benefit of utilizing TFETs in different analog circuit types has been verified recently [45].

These studies have shown that TFETs may contend with low-energy analogue circuits because of the tiny SSs and excellent performance saturation. However, in some analog circuits, the exponential onset of output characteristics, unidirectional current flow, and delayed saturation of output characteristics found in TFETs can be detrimental. The special electrical properties of TFETs, on the other hand, can be used in analog circuits to create new topologies.

The use of TFETs in different forms of analog circuits is currently being studied, as well as the comparative advantages of TFETs over traditional MOSFETs for analog applications [37]. TFETs, like traditional MOSFETs, can be found in amplifier circuits. Si TFETs have recently been compared to a CMOS implementation of the same circuit in traditional source amplifiers. The high-output resistance of the Si TFET [46] gives a greater voltage gain compared to a comparable CMOS amplifier for the Si TFET common-source amplifier.

2.5 TFETS IN CIRCUITS: WHAT THE FUTURE BRINGS

TFETs have obvious advantages over traditional MOSFETs in low-to medium-frequency energy saving. TFETs can also profit from SRAMs and ultra-smooth analog circuit applications. However, several investigations have revealed that the prevailing TFETs are confined to the low voltage of energy consumption over CMOSs. As a result of the current transport process in a TFET, the TFETs present a greater ION than a CMOS inherent disadvantage, which restricts their total operational frequency in TFET-based circuits. In heterogeneous multicore CMOS-TFET circuits, however, TFET can be utilized [47]. The TFET-based cores are designed for low-voltage operations in heterogeneous CMOS-TFET, while the CMOS-based cores are tuned for greater efficiency at higher voltages. TFETs can therefore be utilized to address dark silicone or to underuse computer capacity because of power constraints [48]. TFETs may utilize the power to minimize power utilization and enhance flow in near-threshold voltage circuits. TFETs may be excellent for low-power applications in the future, including one developing discipline, near-threshold computing. At near-threshold computing, TFETs may be a means of addressing power from a system standpoint. TFETs have the potential to be a big player in this space.

2.6 CONCLUSION

In this chapter, the fundamentals of TFETs, along with unit configuration, operating theory, and critical electrical properties of TFETs, were discussed. This chapter also investigated the differences between a standard MOSFET and a TFET in terms of operating theory and electrical characteristics. This chapter highlighted the issues with the TFETs' simple implementation, especially the issue of low ION. The techniques used to improve ION in TFETs are also given. In this chapter, various electrical properties of TFETs are discussed, which have a significant influence on how they're used in different circuits. The use of TFETs in digital circuits, memories, and analog circuits was discussed.

BIBLIOGRAPHY

[1] Ye, P.D., G.D. Wilk, B. Yang, J. Kwo, S.N. Chu, S. Nakahara, H.J. Gossmann, J.P. Mannaerts, M. Hong, K.K. Ng, and J. Bude, "GaAs Metal–Oxide–Semiconductor Field-effect Transistor with Nanometer-thin Dielectric Grown by Atomic Layer Deposition", Applied Physics Letters, Vol. 83, No. 1, PP. 180–182, 7 July 2003.

[2] Ionescu, A.M., L. Lattanzio, G.A. Salvatore, L. De Michielis, K. Boucart, and D. Bouvet, "The Hysteretic Ferroelectric Tunnel FET", IEEE Transactions on Electron Devices, Vol. 57, No. 12, PP. 3518–3524, 28 October 2010.

[3] Kumar, M.J., and S. Janardhanan, "Doping-less Tunnel Field Effect Transistor: Design and Investigation", IEEE Transactions on Electron Devices, Vol. 60, No. 10, PP. 3285–3290, 15 August 2013.

[4] Der Agopian, P.G., J.A. Martino, A. Vandooren, R. Rooyackers, E. Simoen, A. Thean, and C. Claeys, "Study of Line-TFET Analog Performance Comparing with Other TFET and MOSFET Architectures", Solid-State Electronics, Vol. 128, PP. 43–47, 1 February 2017.

[5] Balestra, F. Tunnel FETs for Ultra Low Power Nanoscale Devices, 7 February 2018. Available at: https://www. openscience. fr/IMG/pdf/iste_componano 18v1n.

[6] Dutta, U., M.K. Soni, and M. Pattanaik, "Design and Analysis of Tunnel FET for Low Power High Performance Applications", International Journal of Modern Education & Computer Science, Vol. 10, No. 1, 1 January 2018.

[7] Samal, A., K.P. Pradhan, and S.K. Mohapatra, "Improvising the Switching Ratio Through Low-k/High-k Spacer and Dielectric Gate Stack in 3D FinFET-a Simulation Perspective", Silicon, PP. 1–6, 3 August 2020.

[8] Datta, S., "Recent Advances in High Performance CMOS Transistors: From Planar to Non-Planar", The Electrochemical Society Interface, Vol. 22, No. 1, PP. 41–46, 2013.

[9] Iwai, H., "Future of Logic Nano CMOS Technology", IEEE EDS DL, IIT-Bombay, January 2015.

[10] Saidulu, B., and A. Manoharan, "Power and Area Efficient Opamp for Biomedical Applications Using 20 nm-TFET", in Microelectronics, Electromagnetics and Telecommunications 2018. Springer, Singapore, PP. 207–215.

[11] Singh, S., and B. Raj, "Analytical Modeling and Simulation Analysis of T-shaped III-V Heterojunction Vertical T-FET", Superlattices and Microstructures, Elsevier, Vol. 147, PP. 106717, November 2020.
[12] Chawla, T., M. Khosla, and B. Raj, "Optimization of Double-gate Dual Material GeOI-Vertical TFET for VLSI Circuit Design", IEEE VLSI Circuits and Systems Letter, Vol. 6, No. 2, PP. 13–25, August 2020.
[13] Kaur, M., N. Gupta, S. Kumar, B. Raj, and Arun Kumar Singh, "RF Performance Analysis of Intercalated Graphene Nanoribbon Based Global Level Interconnects", Journal of Computational Electronics, Springer, Vol. 19, PP. 1002–1013, June 2020.
[14] Wadhwa, G., and B. Raj, "An Analytical Modeling of Charge Plasma Based Tunnel Field Effect Transistor with Impacts of Gate Underlap Region", Superlattices and Microstructures, Elsevier, Vol. 142, PP. 106512, June 2020.
[15] Singh, S., and B. Raj, "Modeling and Simulation Analysis of SiGe Hetrojunction Double Gate Vertical T-shaped Tunnel FET", Superlattices and Microstructures, Elsevier, Vol. 142, PP. 106496, June 2020.
[16] Singh, S., and B. Raj, "A 2-D Analytical Surface Potential and Drain Current Modeling of Double-Gate Vertical T-shaped Tunnel FET", Journal of Computational Electronics, Springer, Vol. 19, PP. 1154–1163, April 2020.
[17] Singh, S., S. Bala, B. Raj, and B. Raj, "Improved Sensitivity of Dielectric Modulated Junctionless Transistor for Nanoscale Biosensor Design", Sensor Letter, ASP, Vol. 18, PP. 328–333, April 2020.
[18] Kumar, V., S. Kumar, and B. Raj, "Design and Performance Analysis of ASIC for IoT Applications", Sensor Letter, ASP, Vol. 18, PP. 31–38, January 2020.
[19] Wadhwa, G., and B. Raj, "Design and Performance Analysis of Junctionless TFET Biosensor for High Sensitivity", IEEE Nanotechnology, Vol. 18, PP. 567–574, 2019.
[20] Patel, N., A. Ramesh, and S. Mahapatra, "Drive Current Boosting of n-type Tunnel FET with Strained SiGe Layer at Source", Microelectronics Journal, Vol. 39, PP. 1671–1677, 2008.
[21] Wadhera, T., D. Kakkar, G. Wadhwa, and B. Raj, "Recent Advances and Progress in Development of the Field Effect Transistor Biosensor: A Review", Journal of Electronic Materials, Springer, Vol. 48, No. 12, PP. 7635–7646, December 2019.
[22] Singh, S., and B. Raj, "Design and Analysis of Hetrojunction Vertical T-shaped Tunnel Field Effect Transistor", Journal of Electronics Material, Springer, Vol. 48, No. 10, PP. 6253–6260, October 2019.
[23] Raj, A., S. Singh, K.N. Priyadarshani, R. Arya, and A. Naugarhiya, "Vertically Extended Drain Double Gate S i 1− x G ex Source Tunnel FET: Proposal & Investigation For Optimized Device Performance", Silicon, PP. 1–6, 28 July 2020.
[24] Goyal, C., J.S. Ubhi, and B. Raj, "A Low Leakage CNTFET Based Inexact Full Adder for Low Power Image Processing Applications", International Journal of Circuit Theory and Applications, Wiley, Vol. 47, No. 9, PP. 1446–1458, September 2019.
[25] Sharma, S.K., B. Raj, and M. Khosla, "Enhanced Photosensivity of Highly Spectrum Selective Cylindrical Gate In1-xGaxAs Nanowire MOSFET Photodetector", Modern Physics Letter-B, Vol. 33, No. 12, PP. 1950144, 2019.

[26] Singh, J., and B. Raj, "Design and Investigation of 7T2M NVSARM with Enhanced Stability and Temperature Impact on Store/Restore Energy", IEEE Transactions on Very Large Scale Integration Systems, Vol. 27, No. 6, PP. 1322–1328, June 2019.
[27] Bhardwaj, A.K., S. Gupta, B. Raj, and Amandeep Singh, "Impact of Double Gate Geometry on the Performance of Carbon Nanotube Field Effect Transistor Structures for Low Power Digital Design", Computational and Theoretical Nanoscience, ASP, Vol. 16, PP. 1813–1820, 2019.
[28] Tamak, P., and R. Mehra, "Review on Tunnel Field Effect Transistors (TFET)", International Research Journal of Engineering and Technology, Vol. 4, No. 7, PP. 1195–1200, 2017.
[29] Goyal, C., J. Subhi, and B. Raj, "Low Leakage Zero Ground Noise Nanoscale Full Adder using Source Biasing Technique", Journal of Nanoelectronics and Optoelectronics, American Scientific Publishers, Vol. 14, PP. 360–370, March 2019.
[30] Singh, A., M. Khosla, and B. Raj, "Design and Analysis of Dynamically Configurable Electrostatic Doped Carbon Nanotube Tunnel FET", Microelectronics Journal, Elesvier, Vol. 85, PP. 17–24, March 2019.
[31] Gupta, A., M. Ganeriwala, and N.R. Mohapatra, "An Unified Charge Centroid Model for Silicon and Low Effective Mass III-V Channel Double Gate MOS Transistors", in 2019 32nd International Conference on VLSI Design and 2019 18th International Conference on Embedded Systems (VLSID), 5 January 2019, PP. 163–167.
[32] Lin, P.S., and B.Y. Tsui, "A Comprehensive Evaluation of the Performance of Fin-type Epitaxial Tunnel Layer (ETL) Tunnel FET", in 2017 International Conference on Electron Devices and Solid-State Circuits (EDSSC), 18 October 2017, PP. 1–2.
[33] Xie, H., H. Liu, S. Wang, S. Chen, T. Han, and W. Li, "Improvement of Electrical Performance in Heterostructure Junctionless TFET Based on Dual Material Gate", Applied Sciences, Vol. 10, No. 1, PP. 126, January 2020.
[34] Goyal, C., J.S. Ubhi, and B. Raj, "A Reliable Leakage Reduction Technique for Approximate Full Adder with Reduced Ground Bounce Noise", Journal of Mathematical Problems in Engineering, Hindawi, Vol. 2018, PP. 16, Article ID 3501041, 15 October 2018.
[35] Shokry, F., A. Shaker, M. Elsaid, and M. Abouelatta, "Design of Extended Channel Ge-source TFET for Low Power Applications", International Journal of Integrated Engineering, Vol. 12, No. 8, PP. 191–197, 30 August 2020.
[36] Wadhwa, G., and B. Raj, "Label Free Detection of Biomolecules Using Charge-Plasma-Based Gate Underlap Dielectric Modulated Junctionless TFET", Journal of Electronic Materials (JEMS), Springer, Vol. 47, No. 8, PP. 4683–4693, August 2018.
[37] Settino, F., S. Strangio, M. Lanuzza, F. Crupi, P. Palestri, and D. Esseni, "Simulations and Comparisons of Basic Analog and Digital Circuit Blocks Employing Tunnel FETs and Conventional FinFETs", in 2017 Fifth Berkeley Symposium on Energy Efficient Electronic Systems & Steep Transistors Workshop (E3S), 19 October 2017, PP. 1–3.
[38] Wadhwa, G., and B. Raj, "Parametric Variation Analysis of Charge-Plasma-based Dielectric Modulated JLTFET for Biosensor Application", IEEE Sensor Journal, Vol. 18, No. 15, 1 August 2018.

[39] Yadav, D., S.S. Chouhan, S.K. Vishvakarma, and B. Raj, "Application Specific Microcontroller Design for IoT Based WSN", Sensor Letter, ASP, Vol. 16, PP. 374–385, May 2018.
[40] G. Singh, R.K. Sarin, and B. Raj, "Fault-Tolerant Design and Analysis of Quantum-Dot Cellular Automata Based Circuits", IEEE/IET Circuits, Devices & Systems, Vol. 12, PP. 638–664, 2018.
[41] Singh, J., and B. Raj, "Modeling of Mean Barrier Height Levying Various Image Forces of Metal Insulator Metal Structure to Enhance the Performance of Conductive Filament Based Memristor Model", IEEE Nanotechnology, Vol. 17, No. 2, PP. 268–267, March 2018 (SCI).
[42] Jain, A., S. Sharma, and B. Raj, "Analysis of Triple Metal Surrounding Gate (TM-SG) III-V Nanowire MOSFET for Photosensing Application", Optoelectronics Journal, Elsevier, Vol. 26, No. 2, PP. 141–148, May 2018.
[43] Jain, N., and B. Raj, "Parasitic Capacitance and Resistance Model Development and Optimization of Raised Source/Drain SOI FinFET Structure for Analog Circuit Applications", Journal of Nanoelectronics and Optoelectronins, ASP, USA, Vol. 13, PP. 531–539, April 2018.
[44] Ahmad, S., N. Alam, and M. Hasan, "Robust TFET SRAM Cell for Ultra-low Power IoT Applications", AEU-International Journal of Electronics and Communications, Vol. 89, PP. 70–76, 1 May 2018.
[45] Gupta, A.K., A. Raman, and N. Kumar, "Design and Investigation of a Novel Charge Plasma-based Core-shell Ring-TFET: Analog and Linearity Analysis", IEEE Transactions on Electron Devices, Vol. 66, No. 8, PP. 3506–3512, 9 July 2019.
[46] Strangio, S., F. Settino, P. Palestri, M. Lanuzza, F. Crupi, D. Esseni, and L. Selmi, "Digital and Analog TFET Circuits: Design and Benchmark", Solid-State Electronics, Vol. 146, PP. 50–65, 1 August 2018.
[47] Ionescu, A. "Beyond CMOS: Steep-Slope Devices and Energy Efficient Nanoelectronics", in High Mobility Materials for CMOS Applications 1 January 2018. Woodhead Publishing, Sawston, United Kingdom, PP. 281–305.
[48] Jeyanthi, J.E., and T.S. Arunsamuel, "Heterojunction Tunnel Field Effect Transistors–A Detailed Review", in 2020 5th International Conference on Devices, Circuits and Systems (ICDCS), 5 March 2020, PP. 326–329.

Chapter 3

Trends and Challenges in VLSI Fabrication Technology

Vikas Maheshwari, Neha Gupta, Md Rashid Mahmood, and Sangeeta Jana Mukhopadhyay

CONTENTS

3.1 INTRODUCTION

One of the remarkable features of the semiconductor device that speeds up the dramatic development in the field of information technology is that reducing the size of the device boosts both speed and efficiency while decreasing the price. This is because of the continuous advancements and improvements in the device architecture and the physical downscaling of the devices. Modern complementary metal–oxide semiconductor (CMOS) technology, especially the downscaling of the transistor size, is responsible for this and leads to an exponential increase in the number of transistors on

DOI: 10.1201/9781003311379-3

an integrated circuit. The process for the downscaling of the devices is not only a shrinking of device surface architecture; it also requires the downscaling of the vertical dimensions, as such the source and drain junction depth, gate oxide thickness, and so on. This scaling approach cannot continue as beyond 45-nm technology, the ultrathin gate oxide and ultra-short channel length result in an undesired phenomenon of gate- and drain-to-source leakage currents to an unacceptable level because of tunneling and drain-induced barrier lowering (DIBL). For the prevention of tunneling and leakage currents, physical more thick layers of dielectrics are required. For a thicker gate dielectric, the MOS structure requires a dielectric material with a high dielectric constant to maintain its electrical requirements and performance. CMOS fabrication design principles and processes must be compatible with proposed new high-k dielectric materials and the latest production techniques, as well as with other semiconductor materials utilized in advanced CMOS high-speed integrated circuits. The k-values of the new dielectric material, as well as the alignment of the conduction and valence bands regarding the silicon band structure, are critical in order to minimize leakage current. The scaling of MOS devices must be continued under this environment, despite a decade of study and improved efforts by numerous research organizations using a variety of methodologies. Approaches that have been proposed include the use of strained silicon-on-insulator (sSOI) in place of silicon bulk; the use of materials with high-mobility channel properties, such as germanium, GaAs, graphene, and others; low dielectric constant interconnects; high-k gate dielectric; and some nonplanar CMOS structures, such as fin field-effect transistors (FinFETs) and others.

This chapter deals with the research based on the introduction of the novel material in the fabrication of semiconductor devices. In particular, CMOS fabrication technology is facing difficult times; therefore, novel material must be used to fabricate the transistors. Conventionally, the gate electrode and the gate dielectric of the MOS devices are made of polycrystalline silicon and SiO_2, respectively, as shown in Figure 3.1. Beyond the 45-nm technology node, semiconductor industries returned back to the metal gate technology together with the use of high-dielectric (high-k) materials as a gate dielectric. A compatible high-k dielectric material is not easy to decide as the selection of material must be done on some important parameters, such as it must have a higher resistivity, be thermally stable, and form a perfect interface between silicon and a suitable barrier layer. In 2007, Intel, for the first time, introduced a processor having transistors in which the polysilicon gate electrode and the SiO_2 dielectric layer were replaced by a metal compound or novel metal and a hafnium-based high-k dielectric (HfO_2) material, respectively. Zirconium dioxide (ZrO_2) is also one of the best alternate high-k dielectric gate materials that provide a smaller physical thickness. It is a nanosized material and reduces the direct tunneling off current. But it has been reported

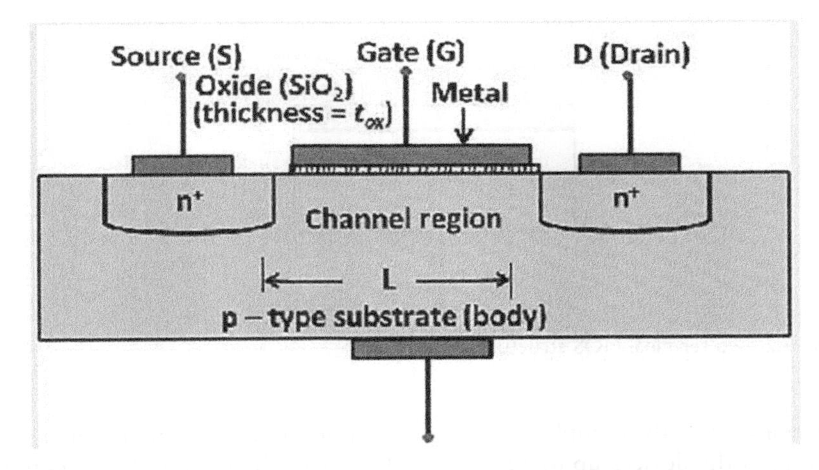

Figure 3.1 Cross section view of metal–oxide–semiconductor field-effect transistor.

to be unstable in contact with silicon. Later, lanthanide oxides are used along with ternary metal oxide compounds such as hafnium-aluminates and hafnium-silicates.

3.2 BACKGROUND

Originally, bipolar technology dominated integrated circuit fabrication technology. But in the late 1970s, MOS technology emerged as one of the most advanced and promising technology for the implementation of very large-scale integration (VLSI) circuits with lower power consumption and higher device-packing density. Figure 3.2 illustrates the three layers of the MOS construction:

- Polysilicon metal gate electrodes
- SiO2-insulating oxide layer
- P/N bulk semiconductors

The gate electrodes of the MOS devices are made of highly doped polycrystalline silicon generally called polysilicon. For last two decades, polysilicon was the preferred material over metal for gate electrodes. The MOS device structure based on the stack of poly-SiO_2-silicon has proved to be extremely successful and has been preferred in the electronics industry for last so many years. Initially, the metal gate was preferred for large operating voltage. Due to aggressive scaling, operating voltage lowered, and manufactures transitioned to using polysilicon as the gate material. Other important reasons for the initial switch to polysilicon are the problems related to fabrication

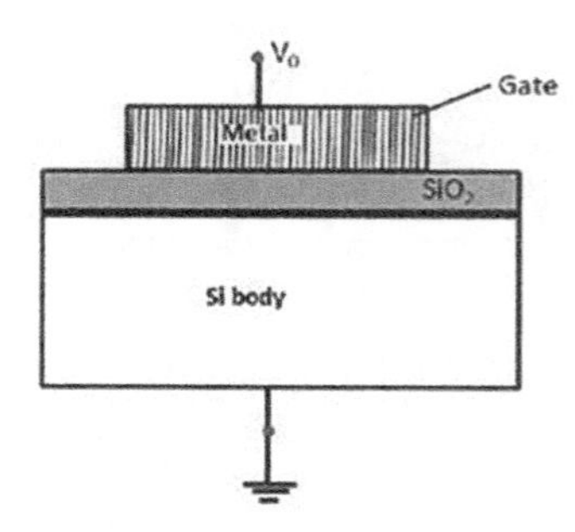

Figure 3.2 Two terminal MOS structure.

processes. After the completion of the initial doping, annealing requires the heating of the devices up to 1200–1400°C. In such high-temperature conditions, metal gates would melt whereas polysilicon gates would not. For the past few decades, the device channel length has been reduced to improve device performance, resulting in a higher level of short channel effects. To counteract these undesirable effects, the gate's physical thickness electrode is reduced to maintain channel control. The SiO_2 layer, or gate dielectric material, is particularly important in determining the performance of MOS devices. In other words, the feasibility of the MOS devices is strongly dependent on gate oxide reliability because the gate dielectric behaves as a capacitor formed between the channel and the gate electrode. In ideal CMOS/MOS devices, the gate dielectric material provides a perfect barrier to the flow of electrons toward the gate electrode from the channel. Selecting the proper material for the gate dielectric is very important as the property of the gate dielectrics determines the gate leakage current because this gate leakage current is exponentially dependent on the barrier height produced by the dielectric layer thickness. The reliability and the operating voltage of the device also depend on the dielectric strength of the material. However, dielectric SiO_2 is not perfect, it also suffers from some measurable degradation because of aggressive scaling down of the size of the MOS devices into very deep submicron (VDSM) regime. For the last decade, the physical thickness of SiO_2 is also reduced in order to fulfill the demand for low-power consumption and high-performance CMOS technology. A very thin layer of gate oxide results in increased level of leakage current toward gate electrodes, called the hot electron effect. The International Technology Roadmap for Semiconductors (ITRS) predicted that beyond 45-nm technology, the amount of the gate leakage current will exceed the subthreshold leakage current for the same oxide layer material. Therefore, insulating the SiO_2 gate oxide must be replaced with the newly introduced high-k dielectric material that allows a thicker oxide layer to lower the gate electrode leakage current. Dielectric material with high-k constant is also important to maintain the capacitance between the gate electrode and the channel without reducing the thickness of the gate oxide layer. Transistor characteristic is also varied as the sizes are scaled down to the nanometer regime. Additionally, newly proposed materials and advanced performance enhancement techniques also are sources for parameter variation. These parameter changes result in a serious threat to the cost-effective utilization of the advanced technology. Scaling down MOS devices beyond the 45-nm dimension size requires improvements, and a lot of research on some important and typical constraints include the following:

1. Charge carriers transferring from the source to the drain and from the drain to the body of MOS devices
2. Charge carriers moving from the channel through the thin gate oxide layer toward gate electrodes
3. Controlling of the doping concentration and the movement of the dopant atoms in the channel and thickness of the source and the drain doped region of the MOS device

Also, as per the observation with scaling, the mobility of the electrons and holes in the deep saturation region is not inadequate to meet the required transistor performance criteria. Scaling the device dimensions horizontally and vertically without scaling the supply voltage produces the electric fields in the device to increase to undesired levels. A classical planer substrate is not able to control adequately short channel effects and statistical variability for very small dimension transistors. The scaling of the MOS devices is also limited by short channel effects by reducing the threshold voltage of the device, resulting in an undesired increment in subthreshold leakage currents. A significant portion of the channel of the short channel devices is depleted by the source, and drain-doped region that reduces the amount of charge in the channel. Therefore, the threshold voltage of the device reduces. In addition to subthreshold leakage current in the MOS devices, because of the high oxide field's significant amount of leakage current is flown from the drain to bulk. For the heavily doped gate to drain overlap region, high fields in the silicon surface cause band bending greater than the silicon band gap over a very short distance. At the surface of the drain, in the depletion region, electrons transferring from the valance band into the conduction band result in a drain-to-bulk current. The high performance of the VLSI circuits is achieved by a transistor scaling process. Each 30% reduction in the scalability of CMOS integrated circuit technology does the following:

1. Reduce gate delays by 30% allows 43% increase in maximum clock frequency
2. Reduce parasitic capacitance by 30%
3. Double the device density
4. Reduce energy and active power consumption per transition by 65% and 50%, respectively

Packaging technology also requires improvements along with the advancement of the CMOS technology scaling with optimized cost and performance levels. System architecture also needs to maximize the performance by improving some factors, such as bandwidth, heat extraction, Input/Output (I/O) density, and power distribution, properly. A perfect balance between a circuit's performance and its power consumption is today's topmost requirement in designing VLSI high-speed devices. Future scaling technology needs foremost advancements in many fields such as

1. improving non-optical exposure and lithography methodology.
2. improving the smaller dimension transistor design to achieve enhanced performance.
3. replacing the silicon-substrate CMOS devices with novel material and structures such as SOI, novel high-k dielectric materials, and strained silicon, among others.
4. scaling the on-chip interconnect wires to improve the circuit delay.
5. more advanced and improved electronic design automation (EDA) tools.

The advancement of the integrated circuit technologies in the sub-quarter-micron regime affected the designing and manufacturing process of the integrated circuits dramatically. Earlier interconnects were considered just as parasitic, but now with the advancement of the technology as complexity of the devices increases, it has become one of the dominant factors that affects high-speed system performance, causing the need for existing methodologies to change accordingly. In a VDSM regime, the intrinsic gate delays, compared to interconnect delays, decrease significantly. In contrast, because of the increased circuit complexity, the worst-case length of the interconnect wire increases up to several miles in large chip size. Therefore, the interconnect wire delay plays an important role in submicron technologies. In the advanced technology era, the integrity of a signal is the one of the critical issues because of ultra-deep submicron effects. Therefore, the efficient and accurate analysis of a large number of interconnect wires is required in advanced and modern integrated circuits.

3.3 REQUIREMENT OF HIGH-K DIELECTRIC MATERIAL

For the last fifty years, the integrating circuit manufacturing industry has been continuously scaling the device size by changing design rules and improving the device and circuit designs dramatically. Therefore, scientists and researchers got the exponential increase in the functionality of the device with high speed, along with a considerable decrement in the power dissipation and cost, within a short span as projected by Moore's law.

Scaling the MOS device size down requires scaling the oxide thickness of the gate dielectric as well. Since 1957, for the manufacturing of the MOS devices, SiO_2 has been the preferred material as a gate oxide layer over others. For the latest technology beyond 45 nm, the oxide layer thickness in MOS devices is predicted to be less than 0.5 nm, which is nearly equivalent to a few atomic layers of SiO_2. Therefore, the next challenging issue is related to the rapid scaling of the dielectric thickness used to provide isolation between the gate and the bulk semiconductor. For an oxide layer thickness less than 0.4 nm, direct tunneling becomes a serious problem, and the problem increases with a decrease in the oxide thickness, where the gate leakage current increases exponentially with a decrease in the dielectric layer thickness. These tunneling currents result in a direct impact on the static power consumption and provide a limiting scenario for the further shrinking of the physical thickness of the oxide dielectric layer. For low-static power consumption, less tunneling current is tolerable, but on the other hand, a thicker dielectric layer is necessary for achieving the required performance of the device.

Over the last few decades, the performance of memory and logic devices has been greatly enhanced by shrinking the transistor size. The capacitance density of MOS devices (C/A) is related to the dielectric constant k and inversely proportional to the dielectric layer thickness, as expressed by

$$C/A = \frac{\epsilon_0 \, k}{t_{ox}}.$$

If the material has high dielectric, k (the relative dielectric constant) and t_{ox} (the material's thickness) are used. Increasing the capacitance density of devices necessitates a substantial reduction in the physical thickness of the SiO_2 layer. A new high-dielectric constant-k-value material for the gate dielectric and metal gate electrodes provides a superior option for reducing tunneling current and degrading gate capacitance. High-k dielectric materials are needed to keep the capacitance density between the gate electrode and the silicon channel without diminishing the SiO_2 layer thickness. These materials have a higher dielectric constant. It is defined as the optimal thickness of SiO_2 required to obtain the same capacitance density as the corresponding oxide thickness of a material with a high dielectric constant (high k), equivalent oxide thickness (EOT) is a function of

$$EOT = \left(\frac{3.9}{k}\right) * t_{ox}.$$

High-k dielectric materials with an EOT of 1.0 nm and low–gate oxide leakage can provide a desired transistor threshold voltage for MOSFETs, as well as transistor channel mobility comparable to SiO_2. Even with a physical thickness (tphy) that is more than the SiO_2 thickness, a high dielectric constant gate oxide layer with a dielectric constant substantially greater than the SiO_2 dielectric constant would attain a smaller equivalent electrical thickness

(t_{eq}) than the SiO_2 layer (t_{ox}). Because the material used must have a greater resistivity than SiO_2, act as a good potential barrier, be thermally stable, and provide an optimal interface with silicon, the SiO_2 layer must be replaced with a dielectric layer with a high dielectric constant. Traditionally, an oxidation process is used to produce a SiO_2 layer over a substrate; however, high dielectric materials need the use of a deposition technique. Atomic layer deposition (ALD) is one of the most promising techniques in the microelectronics production sectors for the deposition of thin films of high dielectric materials over silicon substrates. The thickness of the dielectric layer is regulated at the atomic level in ALD, and the layer structure is coated with a high aspect ratio. The majority of high-k dielectric materials are best deposited via ALD. ALD has emerged as the most advanced technique for fabricating modern CMOS integrated devices that include FinFET structures, wide band-gap semiconductors, and other advanced nanoscale components.

3.4 REQUIRED PROPERTIES OF THE MATERIAL FOR HIGH-K DIELECTRIC

As previously stated, the primary motive for replacing the SiO_2 oxide layer with high-k dielectric materials was to constantly reduce the EOT of MOS devices while retaining a low leakage gate current. It is essential that high-k materials utilized for gate oxide purposes are compatible with the advanced and modern fabrication method, as well as with the other materials used in CMOS circuits. For 45-nm CMOS fabrication technology and beyond, new high-k gate dielectric materials are emerging as an alternate gate oxide layer to standard SiO_2 that can be used to replace SiO_2. The following are the most important requirements:

- Gate leakage current is limited by a high dielectric constant, a large band gap, and a high barrier height.
- It is necessary to have a high amorphous-to-crystalline transition temperature in order to keep the morphology stable following heating.
- It has a high band offset with electrodes when compared to other materials.
- It has high reliability due to the fact that it should not have a high breakdown voltage, no charge trapping, and so on.
- It has a low lattice misfit and a thermal expansion coefficient that is similar to that of silicon.
- High-k bulk films and the interface between high-k and silicon have low defect densities, and the Current-Voltage (C-V) hysteresis is negligible.
- When in contact with a semiconductor substrate, it is thermally and chemically stable.
- It is possible to avoid the formation of a thick low-k interface layer by employing oxygen diffusion coefficients that are low.

Excellent thermodynamic stability on Si, which helps prevent the creation of a low-k SiO_2 interface on the surface. In addition to the requirements mentioned earlier, the gate dielectric material must be thermodynamically stable at the substrate junction as the current modern device fabrication process takes place at a higher temperature. In this situation, the dielectric material should not melt. If a thin layer of high-k dielectric material is not thermodynamically stable at the interface, it starts reacting with the Si at higher temperatures, and a series capacitor will form with the high-k dielectric material layer because of the formation of a low-k interface layer between high-k layer and the silicon substrate. In this condition, the electrical properties of the final high-k gate stack will be dramatically degraded. Mobility degradation is the major issue with the high-k dielectric materials as the channel density of inversion charge carriers is affected by the presence of charge in the dielectric material.

The energy band diagram of the MOS structure with band gap E_g is shown in Figure 3.3. This energy band gap creates a potential barrier height for the tunneling process of the charge carriers. At the metal–oxide interface, this barrier height will be the ϕ_B, that is, the potential difference between the gate and the oxide layer. Similarly, for the electrons traveling from the substrate to the oxide layer at the Si–oxide interface, this barrier height will be ΔE_c, the conduction band offset of the dielectric layers to the Sisubstrate. But in the case of a direct tunneling process, leakage current through the dielectric layer will increase exponentially with a decrease in the barrier height. Leakage current via the gate dielectric is an undesirable phenomenon that must be avoided. Therefore, dielectric material with high-k constant values

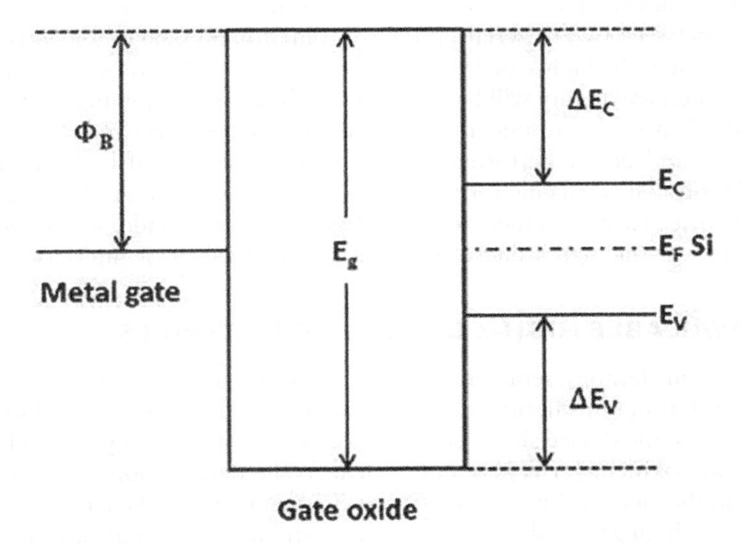

Figure 3.3 Band diagram of the MOS structure.

having a large conduction band offset will be preferred over the other materials. To minimize gate leakage current, an amorphous layer of dielectric is preferred over polycrystalline for gate oxide because amorphous materials don't have the properties of electrical and mass transport along the grain boundaries. Polycrystalline gate dielectric materials have a leaky path because of the grain boundaries in their structure. The grain size and crystalline structure are also varied in the polycrystalline layer, which leads to nonuniform electrical and mechanical properties within the dielectric layer. This causes a reliability issue in integrated circuit performance. The deposition of an amorphous dielectric layer is a great challenge for molecular beam epitaxy (MBE) to be incorporated with the conventional CMOS fabrication process because of low-through output. Commercial ALD equipment is used to deposit a high-quality amorphous dielectric layer over the silicon substrate. In a conventional CMOS fabrication process, devices are heated upto 1000°C for the activation of source/drain and polysilicon dopant after the completion of ion-implantation process. Most of the known metal oxides become crystallizes either after heat treatment or during the deposition process. Therefore, an amorphous dielectric material must have transition temperature of upto 1000°C.

A high-quality interface is to be formed between the high-k dielectric layer and the Si-substrate. All the high-k materials have a high fixed charge density; therefore, they produce a high interface-state density and a considerable shift in flat band voltage. The high-k and Si-substrate interface quality is sometimes determined on the basis of the bonding constraint of these materials. The bonding constraint of any interface is measured in terms of coordination number, that is, the number of atoms available for the covalent bonding between two materials. If the coordination number of a metal oxide is more than 3, the high-k dielectric and Si interface will be over-constrained, and the interface quality will be better. But if the coordination number of the metal oxide is less 3, an under-constrained high-k dielectric and Si interface will form, and device performance will be poor because of high interface-state density. Metal silicide formed at the interface will also grade the device performance. Therefore, metal oxide, as well as metal silicide, should not be available near the high-k dielectric material and Si-substrate interface.

3.5 AVAILABLE HIGH-K DIELECTRIC MATERIALS

For past four decades, semiconductor fabrication industries have been used Si and SiO_2 thin films in integrated circuits. In an advanced CMOS fabrication process, the device size is continuously scaled down, and the thickness of the gate oxide layer is also diminished to the deep nanometer range to maintain the electrical properties of the channel through the gate electrode results in a high gate leakage current. To overcome the limitations imposed by high leakage gate current, dielectric materials with a high dielectric constant were proposed in 2007 as Intel launched integrated circuits with a hafnium-based material as a gate dielectric. These new high-k materials permit a more physical dielectric layer for a considerable reduction in the direct gate leakage current by maintaining a less effective oxide thickness. Therefore, replacing SiO_2 with dielectric materials of highk is biggest technological challenge. Recently metal oxides have been used as dielectric material in MOS devices for gate oxide purposes and stable capacitors in ultra-large-scale integrated circuits (ULSIC). With the replacement of SiO_2 by high-k dielectric materials, gives the same equivalent oxide thickness with a larger physical oxide layer. Various research groups conducted continuous research for the development and improvement of the MOS device performance, several high-k dielectric materials are investigated and proposed for the gate oxide layer but most of them are not having all the required characteristics and properties as per the requirement of the advanced CMOS fabrication process. The electrical performance and physical characteristics of any material are considered to be selected as gate dielectric material, and the list of eligible materials is limited by the design requirements. Finding the suitable replacement of SiO_2 is not an easy task because every material has some challenges. The high-k dielectric material must retain the similar characteristics of SiO_2.

Some of the high-k materials include Y_2O_3, Ta_2O_5, Nb_2O_5, HfO_2, La_2O_3, Al_2O_3, and CeO_2 are under investigation for their role as a dielectric material for the gate oxide layer. All these material films are grown by using ALD, and their dielectric constant ranges from 3.9 to 300. Table 3.1 discusses the some of the popular high-k materials along with their properties. ZrO_2/HfO_2 are popular as the best alternate high-k gate materials that provide a smaller physical thickness, since these are nanosized materials and reduce the direct tunneling off current because of high conduction band offset, low leakage current, and acceptable thermal stability. ZrO_2 has similar characteristics to HfO_2, but as per practical observation, it is found to be unstable in conjunction with Si-substrate. Inspite of so many advantages of using HfO_2 as a suitable candidate for alternative high dielectric constant (high-k) material, it has a major disadvantage of thermal instability. For the perfect stabilization of the Si–metal–oxide interface, the gate oxide must be remained amorphous throughout the CMOS manufacturing process. But unfortunately, HfO_2 layers become crystallize at low temperatures of 450°C when it is deposited by the MBE process and to 530°C when deposited by ALD.

3.5.1 The Challenges for High-k Dielectric Development

The high-k dielectric materials are used to enable continuous scaling of gate oxide thickness and high performance and to control the gate oxide off current in emerging modern high-end nano-electronic MOS devices. The VLSI fabrication industry is facing some major issues regarding the compatibility

Table 3.1 The Properties of Dielectric Materials with High-k Value

Dielectric Material	*Dielectric Constant (k)*	*Energy Band Gap E_g (eV)*	*Conduction Band Offset (ΔE_c)*	*Valence Band Offset (ΔE_v)*
SiO_2	3.9	9	3.5	4.4
Si_3N_4	7.5	5.3	2.2	1.8
Al_2O_3	10	6	3	4.7
$LaAlO_3$	15	5.6	1.6	3.2
ZrO_2	25	5.8	1.4	3.3
HfO_2	25	6	1.5	3.4
TiO_2	40	3.5	1.1	1.3
$BaSrTiO_3$	200–300	–	–	–
Y_2O_3	12–20	5.6	2.3	2.6
La_2O_3	27	4.3	2.3	0.9
Ta_2O_5	20–35	4–4.5	0.3	3.1
CeO_2	26	5.5	–	–
Nb_2O_5	50–250	–	–	–
$SrZrO_3$	180	54	–	–
$HfxSi_{1-x}O_y$	15–25	6	–	–
$SrTiO_3$	300	–	–	–
Gd_2O_5	12–23	5.4	3.2	3.9
Sr_2TiO_4	50	5.2	–	–

of poly-Si gate electrodes with high-k dielectric insulator materials. In many cases, metal oxides with high dielectric constants react with polysilicon, lowering their respective gate oxide dielectric constants. Furthermore, in this situation, controlling the MOSFET threshold voltage is difficult. But in an actual case, it is observed that Fermi-level pinning plays an important role in controlling the MOS threshold voltage. Defect formation at the poly-Si/high-k gate dielectric interface causes Fermi-level pinning in the upper part of the band gap, which causes high-threshold voltage in MOS devices. Figure 3.4 illustrates how the interaction between metal and Si atoms can result in surface dipoles at the poly-Si/high-k interface, which can subsequently affect the interface barrier height and, consequently, the flat-band voltage (VFB) at the interface. In all high-k oxide and poly-Si gate electrode interfaces, an undesirable effective work function (EWF) would be produced, as well as an asymmetric shift in the MOS threshold voltage, because of which the MOS threshold voltage could not be maintained close enough to the mid-gap of Si.

With the introduction of high-k materials, one of the key issues that has arisen is the deterioration of mobility. This effect is caused by charge in the dielectric materials, which reduce the mobility of charge carriers in

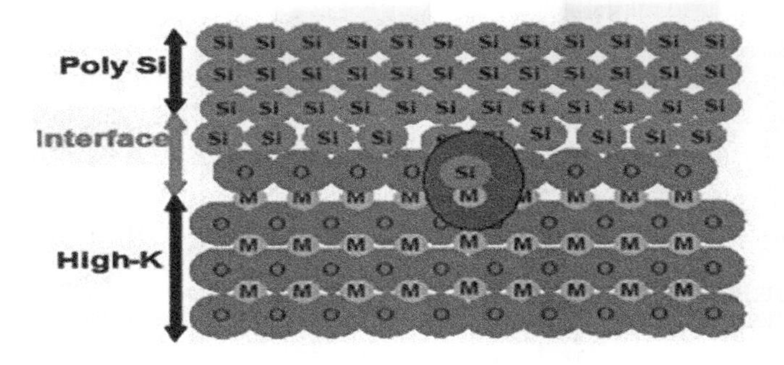

Figure 3.4 Fermi-level pinning phenomena at poly-Si/high-k interface.

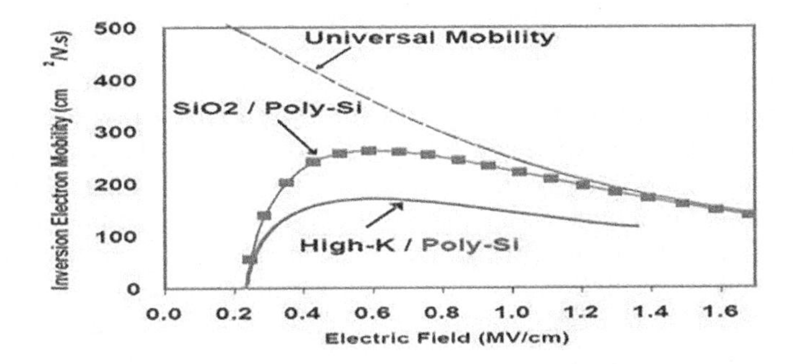

Figure 3.5 Comparing SiO_2/poly-Si and High-k/poly-Si effective electron mobility as a function of effective vertical field.

the channel, resulting in a reduction in the density of the inversion charge carriers in the channel. However, because of the reduced mobility of high-k materials, the advantages of adding high-k materials are discovered to have vanished. This is because all high-k materials contain a large number of charges and traps at the interface. The increase in remote phonon scattering in high-k materials is caused by the ionic polarization, which results in a reduction in mobility. The increased ionic polarization properties of high-k dielectric materials are thought to be one of the contributing factors to their high-k value. As a result, it is possible that the usage of high-k dielectric materials will result in an increase in remote phonon scattering. According to Figure 3.5, the combination of high-k dielectric materials and polysilicon gate electrodes is unsuitable for high-performance devices because the combination of high-k and polysilicon gate electrodes has relatively high threshold voltages and degraded channel mobility, indicating that the combination is not suitable for high-performance devices.

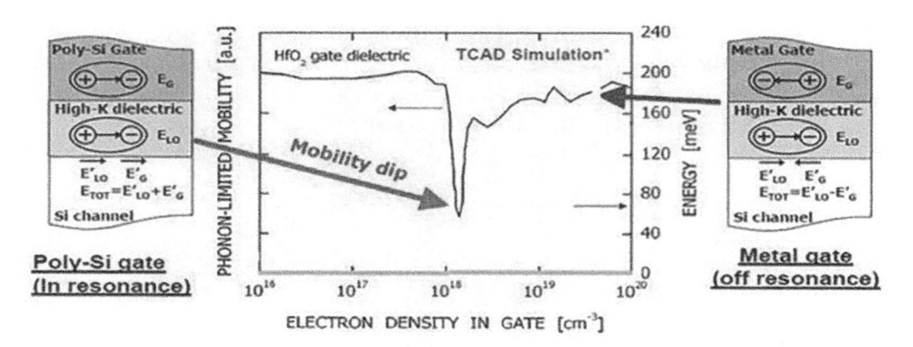

Figure 3.6 Phenomena of surface phonon scattering.

The principal source of channel mobility loss in high-k dielectric materials is surface phonon scattering, as proved both empirically and theoretically. The polarization of metal-oxygen bonds in high-dielectric material layers produces low-energy optical phonons. As seen in Figure 3.6, these phonons can be modeled as dipoles that fluctuate in the channel. The oscillating dipoles are thought to be related to channel electrons in the resonance and gate plasma oscillation circumstances. When the gate carrier density is roughly 1×10^{18} cm^{-3} for a conventional doped polysilicon gate, the resonance occurs. In this case, the appropriate gate plasma energy is contained inside the dominating energy modes of the high-k dielectric materials, namely, the longitudinal and transverse optical modes.

This resonance condition results in a significant loss of mobility. The resonance condition, on the other hand, is not satisfied when the carrier density of the metal gate electrode surpasses 1×10^{20} cm^{-3}. This off-resonance condition weakens the carrier phonon coupling and results in a significant degree of carrier mobility recovery.

Due to the previously mentioned issues, during the fabrication of MOS devices, an ultra-thin SiO_2 layer is purposely grown in between the Si-channel and the high-k dielectric material to minimize the mobility degradation. An additional layer of SiO_2 that will limit the thickness of the high-k dielectric material by increasing the EOT may be considered a trade-off between mobility enhancement and the gate leakage current. With the advancement of the technology beyond the 45-nm node, the thickness of an additional SiO_2 layer must be small as much as possible because of the EOT requirement near about 10 μm. Therefore, the deposition and thickness control of an additional SiO_2 layer over the Si wafer is very crucial. At present, so many advanced deposition techniques are developed, but still, it is a technical challenge to achieve thinner EOT. Because of the replacement of SiO_2 with high-k dielectric material, device mobility is observed to be degraded as a problematic criterion for low EOT applications. Therefore, it is found that polysilicon gate electrodes are not compatible with a high-k dielectric layer, and a metal gate has several advantages over polysilicon gate with high-k dielectric layer. So, poly-Si is being replaced by a metal to make it compatible with high-k material.

3.6 METAL GATE ELECTRODES

To overcome the previously mentioned issues with a doped poly-Si gate electrode is to use metal or metal compound gate electrodes. Metal gate electrodes have a number of advantages over doped polysilicon gate electrodes in modern technology. In order to properly replace severely doped polysilicon gates with metal gate electrodes, metal systems with the proper band-edge work functions must be used. One of the most challenging difficulties is how to manage the work function of the metal gate electrodes once they have been subjected to CMOS fabrication. Two approaches are proposed for the implementation of a metal gate. A single metal electrode is used in the first procedure, and its work function is regulated by integrating extra materials to get it close to the mid-gap to about 4.6eV. In the second technique, dual–metal gate electrodes are implemented with the first metal having a work function near the conduction band of the Si-substrate for NMOS and the second metal having a work function near the valence band of the Si-substrate for PMOS, as seen in Figure 3.7. As a result of the high-k dielectric materials and surrounding materials utilized in the process, the metal gate electrodes must be thermally, mechanically, and chemically stable when used in the CMOS fabrication process. The

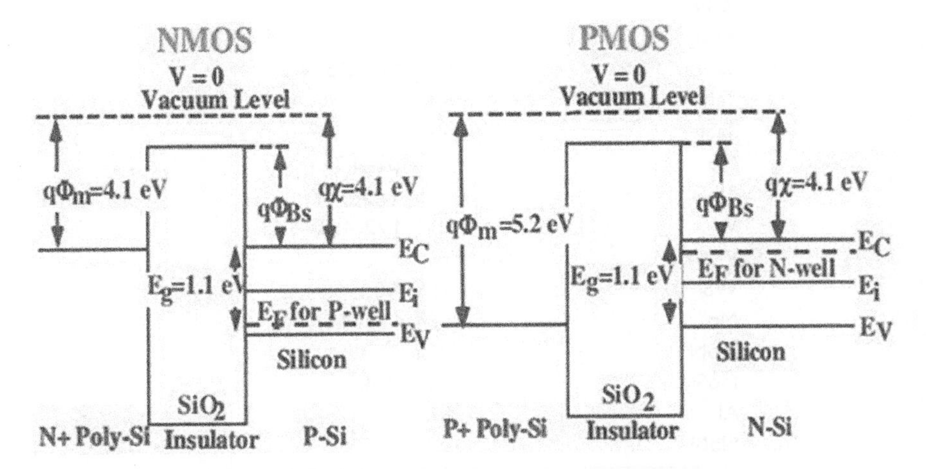

Figure 3.7 Metal gate electrode implementation in NMOS and PMOS devices.

sheet resistance of metal gate electrodes must be kept to a minimum so that they are compatible with CMOS fabrication processes, whether they are used in a gate-first or gate-last approach. It is estimated that more than 1,000 research groups across the world are working on developing a way to precisely quantify the work function of various metals and metal combinations. Traditional CMOS devices substitute poly-Si for a metal with the appropriate set of work functions, such as strongly doped germanium used in PMOS transistors.

p-type poly-Si is used as gate electrodes having a work function about 5.2 eV whereas in NMOS transistors, heavily doped n-type poly-Si is used as gate electrodes having a work function about 4.1 eV. Therefore, it is the primary requirement that the substituting metals or metal compounds should have matching work functions for a better performance of the MOS transistors. The work function values for certain prospective metal gate electrode candidates are shown in Table 3.2. For the designing of a high-performance device, the work function of the materials used for the gate electrode is the important criterion. In polysilicon gate technology, the range of gate work function is limited to values close to the valence band and the conduction band of the Si, but on one hand, in metal gate technology, selecting a suitable

Table 3.2 Various Gate Materials with Work Functions

Gate Material	*Work Function (MOS) in eV*
Ti	4.17, 4.33, 4.6
TiN	4.95
$TiSi_2$	3.67–4.25
Zr	4.05
ZrN	4.6
$ZrSi_2$	–
Ta	4.25, 4.6, 4.15–4.25
TaN	5.41
$TaSi_2$	4.15
Nb	4.3, 4.02–4.3
$NbSi_2$	4.35–4.53
W	4.75, 4.72, 4.55–4.63
WNx	5
WSi_2	4.55–4.8
Mo	4.64, 4.43–4.6
MoN	5.33
$MoSi_2$	4.6–4.8, 4.9
Al	4.1
NbN	–

work function is possible so that device can be redesigned to achieve the best combination of the channel doping and work function. Silicon's conduction and valence bands are quite close to the majority of gate work function values. A metal gate electrode with a mid-gap work function has either an overly high-threshold voltage for high-performance applications or a short-circuit work function.

Metal work functions, on the other hand, are principally responsible for regulating the threshold voltage in dual-gate FET devices. The work function of a metal gate electrode measured in a vacuum differs from the work function value when the metal gate electrode comes into contact with a dielectric substance, complicating matters further. A dipole formation at the metal–dielectric interface alters the overall EWF of a metal–dielectric material combination. The gate dielectric material determines the sort of metal electrode that can be used in certain circumstances. The replacement gate technology is a sophisticated and promising technique for the integration of metal gates by employing a dummy gate material to build the self-aligned gate to the source and gate to the drain structure. This fake gate material is removed using an etching technique similar to polysilicon gate technology and replaced with the desired gate dielectric material and the metal gate electrode. In the CMOS device fabrication process, the integration of NMOS and PMOS transistors remains a challenge as two different work functions are employed, respectively, for NMOS and PMOS transistors. Care should be taken for two steps as

- separate deposition of the metals should be done for NMOS and PMOS.
- the connection of PMOS and NMOS is done in a compact manner by strapping the two different metals.

Avoiding these two conditions and finding a simple approach to adjust the metal's work function would be preferable. Regarding the previously mentioned issues, some of the research groups revealed two ways. The first method involves forming a single molybdenum metal layer and changing the work function by ion implantation of nitrogen into the metal. Because the completed process flow information is not obvious, and the ion implantation procedure only requires a single metal deposition and a photoresist mask, this approach is unreliable. The desired work function is obtained in a second way by intermixing metals. In this process, two metals A and B are deposited on the gate dielectric layer and then selectively etched from the top metal while leaving the bottom metal where it is. Then, the two metals A and B are heated and annealed together. In contrast to the region where only metal A exits as shown in Figure 3.8, at high temperatures, either diffusion or intermixing occurs in between the materials, resulting in complementary

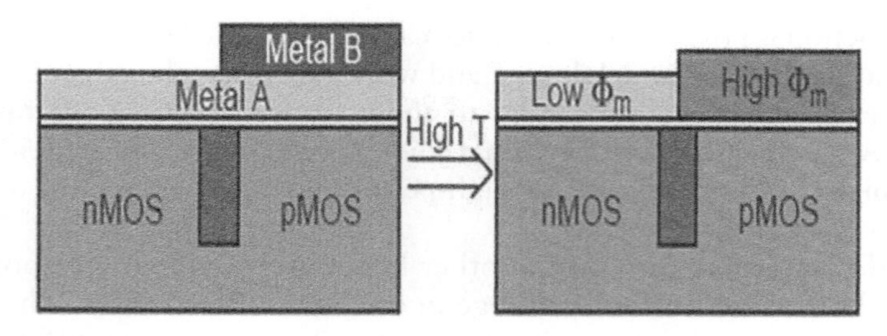

Figure 3.8 Schematic picture of a possible complementary metal gate formation.

work functions. For example, some already known processes can be discussed here:

- Over-stoichiometric process of TiN1+x on Ta or Mo, whereby nitrogen from TiN1+x is diffused into Ta or Mo.
- Process of Ti and Ni intermixing after annealing process; Ti is responsible for the determining of threshold voltage in NMOS transistors and Ni and Ni-rich alloy of Ti determines the threshold voltage for the PMOS transistors.
- Nickel silicidation of doped polysilicon, whereby the doping type and concentration of polysilicon determines the work function and the silicidation hinders the gate electrode depletion effects.

3.6.1 Metal Gate Work Function Discussion

The variation of the metal work function is dependent on the gradient of the nitrogen concentration in the metal film thickness when the metal specimen is left for temperature annealing after sputter deposition of TiN, TiNx, and ZrNx samples in a nitrogenous environment. After annealing the work function of the reactively sputter-deposited TiNx and ZrNx films increases at intermediate gas flows. It is experimentally observed that the thin layer of ZrNx films behaves like a metallic nature whereas a thick layer film is more compound depending on the deposition time. Because the deposition time was the only difference between the samples, it's possible that removing the shutter causes a change in the plasma, necessitating a short period before returning to a steady state. A similar procedure was utilized in the PVD TiN research, and this transitory event could also apply to them: pre-sputter in Ar, pre-sputter in N2 + Ar, and then open the shutter with the substrate directly beneath the plasma. Nitrogen diffusion to the metal/SiO_2 contact during subsequent annealing may increase the work function. Pre-sputtering the target in both metallic and compound modes for many minutes is used to deposit the material. Because of the lowering of nitrogen concentration via diffusion mechanism during the first annealing phases, the retrieved value of work function remains lower in the case of metallic interface behavior. More nitrogen gradient at the metal

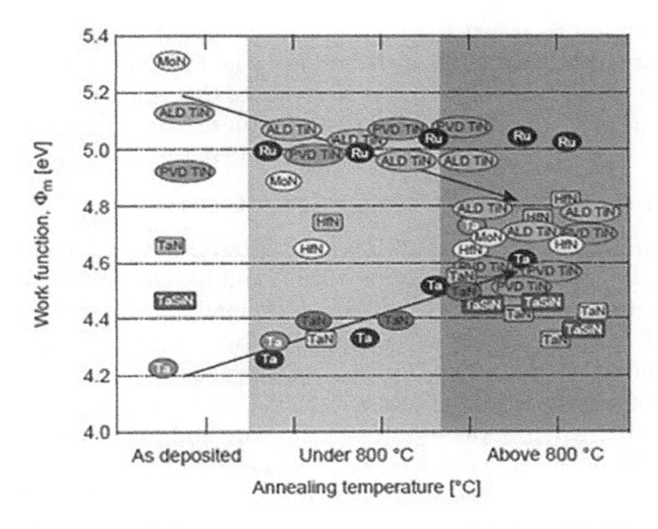

Figure 3.9 As a function of annealing temperature, the work function of various metal gate technologies.

gate and dielectric layer contact is used to derive a larger work function value. This is the same decreased work function scenario reported after annealing at temperatures above 8000°C in many metal gates with SiO_2 dielectric. On SiO_2 dielectric material, all metal gates except Ru have a work function in the mid-gap region, but metal gates on HfO_2 dielectric material have a work function that is substantially lower. The work function of HfN remains constant near the mid-gap before and after annealing. Figure 3.9 depicts a graphical representation of the variance in work function of several materials.

Due to the presence of a large density of extrinsic interface states, Fermi-level pinning occurs, resulting in defects such as vacancies, variations in chemical bonds, and changes in the interface structure. Lowering of the extracted work function of the metal gate electrode is occurred because of the negatively charged extrinsic states below the metal Fermi level. Metals electrodes on HfO_2 dielectric exhibit a lower value of work function is due to either no fermi level pinning or as pinning to a lower level. The work function of a metal intact to a dielectric material is determined by an internal photoemission (IPE) process. It is necessary to expose a semitransparent capacitor construction with extremely thin metal electrodes to ultraviolet light in order to detect the photocurrent. When a negative gate voltage is provided, electrons from the metal are stimulated into the oxide, and the metal/dielectric interface barrier height is removed from the metal. Specifically, the semiconductor/oxide barrier is determined in the reverse polarity scenario. When compared to traditional capacitors, the functioning principle of capacitors utilized in the IPE process is significantly different. The metal used in IPE capacitors is extremely thin, with a thickness of roughly 10 nm, making it almost as transparent to light as a sheet of paper. Because of the low quantity of leakage, thick dielectric material layers are chosen, and as a result, little photocurrent dominates. Because of this, greater sizes of the patterned metal plates are required in order to achieve high photocurrent values.

3.6.2 Issues with Metal Gate Device Integration

The deposition of NMOS and PMOS gate electrodes is problematic because the metal gate materials required for NMOs and PMOS are different. Important gate device parameters, such as gate oxide integrity, threshold voltage stability, work function, and others, are affected by contaminations and plasma damage due to CVD and PVD processes, respectively. The etching process of metal oxides is a critical process as metal gates are having poor oxidation resistance. MOS devices with a single gate and a work function value in the middle of the gap are not able to produce low threshold voltage and good performance, but they are beneficial in fully depleted SOI applications. Devices having dual metal gate electrodes having work function similar to heavily doped polysilicon is a technologically challenging task. Selected metal gate material should be thermodynamically and mechanically stable. There are benefits and drawbacks to both gate-first and gate-last integration systems. The integration technology is chosen primarily on the basis of cost, performance, and yield.

3.6.2.1 Gate-First CMOS Integration Technology

The gate material can be created using a common technique known as "gate first integration technology". A Nitrid/W/TiN gate electrode stack on the top of gate oxide is shown in Figure 3.10. As metal gates are having poor oxidation resistance, therefore, gate materials must be protected from oxidation and wet chemicals. This integration technology is involved in high-temperature steps. In these high-temperature conditions, gate metal electrodes must be thermodynamically stable with their surrounding materials used in the CMOS process. Because of these critical requirements, the selection of metal gate materials and corresponding alternative high-k dielectric materials is limited [56].

3.6.2.2 Integration of Gate-Last CMOS

This integration technology has many advantages over gate-first integration. This is the latest integration technology that gives a promising solution for the problems associated with gate-first integration, such as thermal or chemical stability concerns, the requirement of etching new materials, and stress-induced diffusion issues. After the fabrication of the conventional MOS transistors, a thin gate oxide is deposited, and the top surface is flattened using chemical-mechanical polishing (CMP) technique. The polysilicon gate is etched by using a suitable etching process followed by deposition

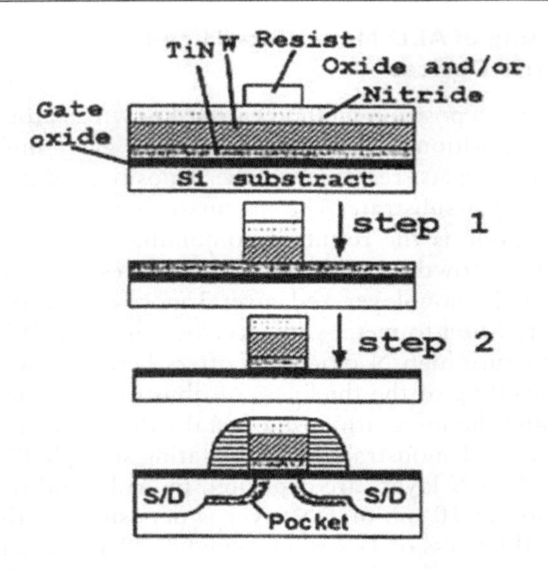

Figure 3.10 Successive etching process involved in convention gate-first integration.

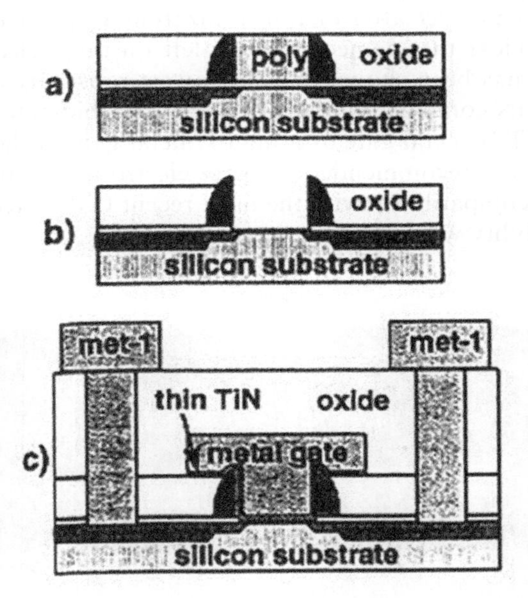

Figure 3.11 Successive etching process involved in convention gate-last integration.

of new gate dielectric material and the metal gate electrode. The final structure of the device is obtained after second patterning by using another CMP process. The complete process is illustrated in Figure 3.11.

3.6.3 Processing of ALD Metal Gate/High-k Dielectric Devices

ALD is a chemical deposition method earlier known as atomic layer epitaxy in which deposition is done by pulsing two gases into the sample chamber. This gas reacts, resulting in the deposition of a thin layer of the desired film on a substrate. The precursor is the name given to the primary gas. A cycle is the result of combining these two gas pulses. During the initial growth on a surface, ALD takes a number of cycles to generate a single monolayer and several more cycles to form a fully closed film. In addition to metal gate electrode films, ALD is widely utilized to manufacture high-N oxides. It offers better uniformity over a large area, controlling of the thickness of films on an atomic level. The process, including the integration of a metal gate, channel strain, and a high-N dielectric, is demonstrated by fabricating suitable PMOS transistors. Different High-N layers are deposited by ALD, and work function is settled by another 10 nm of TiN layer is deposited by the same process. A 180-nm thickness of TiN is deposited by PVD for gate contact. A hard mask patterning is used to etch the gate fingers in an ICP chamber using Cl_2/BCl_3 plasma. Figures 3.12 and 3.13 show a typical device after metal fetch and a device after final metallization, respectively. From Figure 3.12, it is clear that some residue is left on the dielectric after the dry etching. It has been shown that the lowest resistivity of TiN film is achieved at a TiN composition that is close to stoichiometric. The PMOS devices used a TiN metal gate as a high-k/metal gate in the initial iteration. Metal TiN is recommended for gate electrodes because of its low resistivity and compatibility with the most recent CMOS technology and its thermal stability when it is in contact with HfO_2.

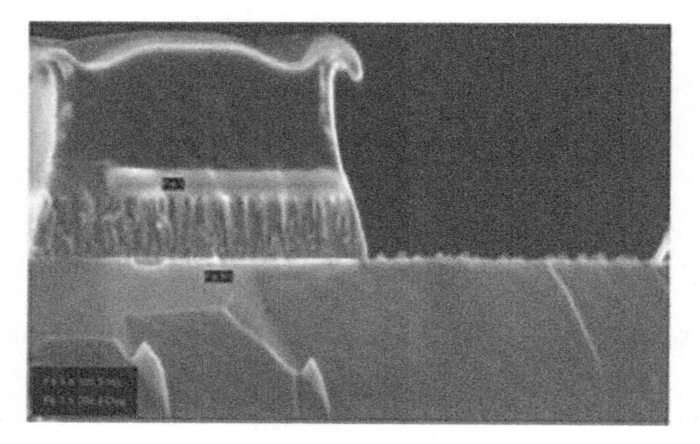

Figure 3.12 TiNGate finger sidewall after dry etch.

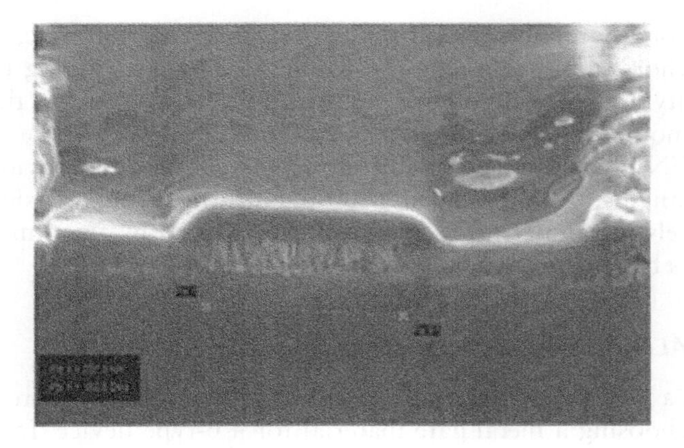

Figure 3.13 Typical device after metallization.

3.6.3.1 ALD of P-Type Metal Gate

Among the most frequently seen options for the p-type metal gate used in transistors are the refractory elements ruthenium (Ru), platinum (Pt), and tungsten (W), all of which are refractory elements (W). Despite the fact that all three of these metals are among a limited number of metals that can be deposited, the ALD method is capable of depositing them all. Tungsten ALD has shown to be the most successful of the other elemental metals tested. While being used in ALD on SiO_2, WF_6, and Si_2H_6 enable a linear W development at temperatures less than 350°C without requiring any additional time for incubation. While the entire mechanism of W deposition is a self-limited reaction, it includes the elimination of the consequent SiHyF contents by the subsequent WF_6 exposure, as well as the stripping of fluorine from WFX species by Si_2H_6, which is a self-limiting reaction. There have been multiple approaches to the ALD process for W that have been researched and demonstrated by various research organizations. For the most part, the researchers focused their efforts on the application of CVD as a seed layer for W contact plugs. The ALD-W-on-TiN-film team created the ALD-W-on-TiN film, which combines SiH_4 and B_2H_6 as metals to bridge the high aspect-ratio gap in the replacement gate. The ALD-W-on-TiN-film team also produced the ALD-W-on-TiN film. When it comes to gap-filling capabilities and electrical performance, MOS capacitors and PMOSFETs outperform their rivals by a wide margin. In an investigation into the electrical properties of SiH4 and B_2H_6, it was found that the former, which is crystalline and has a tensile stress of approximately 2.4 Gpa, may have superior electrical capabilities to the latter. An ALD process was utilized to build polycrystalline TiN on glass at 5000°C, utilizing $TiCl_4$ as a precursor and NH_3 as a reactant, with a deposition rate of less than 0.02

nm/cycle, or 0.1 monolayers in the (111) direction each cycle, according to the authors' findings. Using the Rutherford backscattering technique, the quantity of hydrogen impurity in the resultant layers is less than 0.4%, and the amount of chlorine is less than the detection limit of the technique. In these TiN films, a trace amount of oxygen can be found, with concentrations ranging from 3–5% in 1400-micron-thick films to significantly higher levels in thinner films. This could be caused by the film oxidizing after it has been deposited.

3.6.3.2 ALD of n-Type Metal Gate

Choosing a metal gate material for an n-type device is far more challenging than choosing a metal gate material for a p-type device. In 2002, for the first time, a device with dual metal gates was presented, which was manufactured using Ti as the n-type metal gate and Mo as the p-metal gate, with SiOxNy as the dielectric. The suggested device demonstrated adequate channel mobility, low gate leakage, and no gate depletion. Sputtering techniques are used to deposit a Tin/Ti double layer on SiOxNy, yielding a predicted work function of 4.36 eV for metal Ti. Later, the first-order sensitivity index (FOSI) technique was proposed, which uses full silicidation to produce the polysilicon gate electrode for dual-gate applications. For PMOS and NMOS, a polysilicon layer is first formed on SiON and doped with B and As, respectively, in the FOSI process. The poly-Si was then annealed to cause silicidation after a sufficiently thick Ni layer was grown on it. With Ru as the p-type metal, TaC as the n-type metal, and HfO_2 as the dielectric material, it was demonstrated that satisfactory device behavior could be achieved with dual metal gate integration. As previously noted, TiAl alloy was employed as an n-type MOSFET in the first industrial high-k with metal gate technology at the 45-nm node, which was suggested in 2007. Later in 2015, researchers demonstrated thermal ALD of TiAlC on an ALD HfO_2 surface using TiCl4 as a Ti precursor and trimethyl-aluminum (TMA) as an Al precursor. Temperatures ranging from 300–400°C were used for the $TiCl_4$ pulses, N_2 purges, TMA pulses, and N2 purges, which were all completed in one deposition cycle. It was determined that TiAlC was the predominant component of the film deposited at 400°C, with 55% carbon, 35% titanium, 8% aluminum, and a trace of chlorine as other constituents. Observations of the linear growth curve revealed no sign of incubation. In accordance with AFM results, the layer was amorphous, had low roughness (0.333 nm), and had low resistance. In the C-V measurement, the MOSFET Capacitance (MOSCAP) with a structure of W/TiN/TiAlC/HfO_2/SiO_2/Si showed no hysteresis, indicating that the defect density was minimal in this structure. The EWF of TiAlC layers of varying thicknesses ranged from 4.49–4.79 eV for different thicknesses.

3.7 CONCLUSION

The chapter started with the discussion regarding the introduction to the basic structure of the MOS device and the associated technical problems of the present structure with the latest technologies. As the technology advances, the MOS structures are gone through dramatic transition from the fundamental structure, having poly-Si/SiO_2 structures, to the most advanced structure, having high-k/metal gate structures. In advanced MOS devices, metal gates require ALD process because of the technological deposition challenges. Challenges associated with the various parameters of the MOS device are also discussed in detail. In the future, the complete structure of the MOS device will be a transformed structure that will not use any conventional structure.

REFERENCES

[1] Julius Edgar, L., "Method and Apparatus for Controlling Electric Current", U.S. Patent No. 1,745,175, 28 January 1930.

[2] Kahng, D., "Electric Field Controlled Semiconductor Device", U.S. Patent No. 3,102,230, 27 August 1963.

[3] Wanlass, F., "Low Stand-By Power Complementary Field Effect Circuitry", U.S. Patent No. 3,356,858, 5 December 1967.

[4] Wilk, G.D., R.M. Wallace, and J.M. Anthony, "High-k Gate Dielectrics: Current Status and Materials Properties Considerations", Journal of Applied Physics, Vol. 89, P. 5243, 2001.

[5] Degraeve, R., G. Groeseneken, R. Bellens, J.L. Ogier, M. Depas, P.J. Roussel, and H.E. Maes, "New Insights in the Relation Between Electron Trap Generation and the Statistical Properties of Oxide Breakdown", IEEE Transactions on Electron Devices, Vol. 45, PP. 904–911, 1998.

[6] Pfiester, J.R., F.K. Baker, T.C. Mele, H. Tseng, P.J. Tobin, J.D. Hayden, J.W. Miller, C.D. Gunderson, and L.C. Parrillo, "The Effects of Boron Penetration on P+ Polysilicon Gated PMOS Devices", IEEE Transactions on Electron Devices, Vol. 37, PP. 1842–1851, 1990.

[7] Hattangady, S.V., R. Kraft, D.T. Grider, M.A. Douglas, G.A. Brown, P.A. Tiner, J.W. Kuehne, P.E. Nicollian, and M.F. Pas, "Ultrathin Nitrogen-profile Engineered Gate Dielectric Films", in Proceedings of the International Electron Devices Meeting, San Francisco, CA, USA, 8–11 December 1996, PP. 495–498.

[8] Wu, Y., and G. Lucovsky, "Ultrathin Nitride/Oxide (N/O) Gate Dielectrics for P+ Polysilicon Gated PMOSFETs Prepared by a Combined Remote Plasma Enhanced CVD/Thermal Oxidation Process", IEEE Electron Device Letters, Vol. 19, PP. 367–369, 1998.

[9] Wang, X.W., Y. Shi, T.P. Ma, G.J. Cui, T. Tamagawa, J.W. Golz, B.L. Halpen, and J.J. Schmitt, "Extending Gate Dielectric Scaling Limit by Use of Nitride or Oxynitride", in Proceedings of the 1995 Symposium on VLSI Technology, Kyoto, Japan, 6–8 June 1995, PP. 109–110.

[10] Yang, H., and G. Lucovsky, "Integration of Ultrathin (1.6/spl sim/2.0 nm) RPECVD Oxynitride Gate Dielectrics into Dual Poly-Si Gate Submicron

CMOSFETs", in Proceedings of the International Electron Devices Meeting, Washington, DC, USA, 5–8 December 1999, PP. 245–248.
[11] Ellis, K.A., and R.A. Buhrman, "Time-dependent Diffusivity of Boron in Silicon Oxide and Oxynitride", Applied Physics Letters, Vol. 74, PP. 967–969, 1999.
[12] Ellis, K.A., and R.A. Buhrman, "Boron Diffusion in Silicon Oxides and Oxynitrides", Journal of the Electrochemical Society, Vol. 145, PP. 2068–2074, 1998.
[13] Lucovsky, G., Y. Wu, H. Niimi, V. Misra, and J.C. Phillips, "Bonding Constraints and Defect Formation at Interfaces Between Crystalline Silicon and Advanced Single Layer and Composite Gate Dielectrics", Applied Physics Letters, Vol. 74, PP. 2005–2007, 1999.
[14] Seungheon, S., W.S. Kim, J.S. Lee, T.H. Choe, J.H. Choi, M.S. Kang, U.I. Chung, N.I. Lee, K. Fujihara, H.K. Kang, et al., "Design of Sub-100 nm CMOSFETs: Gate Dielectrics and Channel Engineering", in Proceedings of the the 2000 Symposium on VLSI Technology, Honolulu, HI, USA, 13–15 June 2000, PP. 190–191.
[15] Robertson, J., "Band Offsets of Wide-band-gap Oxides and Implications for Future Electronic Devices", Journal of Vacuum Science & Technology B, Vol. 18, PP. 1785–1791, 2000.
[16] Brar, B., G.D. Wilk, and A.C. Seabaugh, "Direct Extraction of the Electron Tunneling Effective Mass in Ultrathin SiO2", Applied Physics Letters, Vol. 69, PP. 2728–2730, 1996.
[17] Vogel, E.M., K.Z. Ahmed, B. Hornung, W.K. Henson, P.K. McLarty, G. Lucovsky, J.R. Hauser, and J.J. Wortman, "Modeled Tunnel Currents for High Dielectric Constant Dielectrics", IEEE Transactions on Electron Devices, Vol. 45, PP. 1350–1355, 1998.
[18] Pillai, K.P.P., "Fringing Field of Finite Parallel-plate Capacitors", Proceedings of the Institution of Electrical Engineers, Vol. 117, PP. 1201–1204, 1970.
[19] Ma, T.P., "Making Silicon Nitride Film a Viable Gate Dielectric", IEEE Transactions on Electron Devices, Vol. 45, PP. 680–690, 1998.
[20] Dey, S.K., and J.J. Lee, "Cubic Paraelectric (Nonferroelectric) Perovskite PLT Thin Films with High Permittivity for ULSI DRAMs and Decoupling Capacitors", IEEE Transactions on Electron Devices, Vol. 39, PP. 1607–1613, 1992.
[21] Takeuchi, H., and T.-J. King, "Scaling Limits of Hafnium–silicate Films for Gate-Dielectric Applications", Applied Physics Letters, Vol. 83, PP. 788–790, 2003.
[22] Seong, N.-J., S.-G. Yoon, S.-J. Yeom, H.-K. Woo, D.-S. Kil, J.-S. Roh, and H.-C. Sohn, "Effect of Nitrogen Incorporation on Improvement of Leakage Properties in High-k HfO2 Capacitors Treated by N2-plasma", Applied Physics Letters, Vol. 87, P. 132903, 2005.
[23] Zhao, C., T. Witters, B. Brijs, H. Bender, O. Richard, M. Caymax, T. Heeg, J. Schubert, V.V. Afanas'ev, A. Stesmans, et al., "Ternary Rare-earth Metal Oxide High-k Layers on Silicon Oxide", Applied Physics Letters, Vol. 86, P. 132903, 2005.
[24] Barlage, D., R. Arghavani, G. Dewey, M. Doczy, B. Doyle, J. Kavalieros, A. Murthy, B. Roberds, P. Stokley, and R. Chau, "High-frequency Response of 100 nm Integrated CMOS Transistors with High-K Gate Dielectrics", in

Proceedings of the IEEE International Electron Devices Meeting, Washington, DC, USA, 2–5 December 2001, PP. 10.16.11–10.16.14.

[25] Mistry, K., C. Allen, C. Auth, B. Beattie, D. Bergstrom, M. Bost, M. Brazier, M. Buehler, A. Cappellani, R. Chau, et al., "A 45nm Logic Technology with High-k+Metal Gate Transistors, Strained Silicon, 9 Cu Interconnect Layers, 193nm Dry Patterning, and 100% Pb-free Packaging", in Proceedings of the IEEE International Electron Devices Meeting, Washington, DC, USA, 10–12 December 2007, PP. 247–250.

[26] Choi, J.H., Y. Mao, and J.P. Chang, "Development of Hafnium Based High-k Materials—A Review", Materials Science and Engineering R: Reports, Vol. 72, PP. 97–136, 2011.

[27] Huang, C., N.D. Arora, A.I. Nasr, and D.A. Bell, "Effect of Polysilicon Depletion on MOSFET I-V Characteristics", Electronics Letters, Vol. 29, PP. 1208–1209, 1993.

[28] Misra, V., "Dual Metal Gate Selection Issues", in Proceedings of the 6th Annual Topical Research Conference on Reliability, Austin, TX, USA, 27–28 October 2003.

[29] Hauser, J.R., and K. Ahmed, "Characterization of Ultra-thin Oxides Using Electrical C-V and I-V Measurements", AIP Conference Proceedings, Vol. 449, PP. 235–239, 1998.

[30] Brown, G.A., P.M. Zeitzoff, G. Bersuker, and H.R. Huff, "Scaling CMOS: Materials & Devices", Materials Today, Vol. 7, PP. 20–25, 2004.

[31] Qiang, L., R. Lin, P. Ranade, K. Tsu-Jae, and H. Chenming, "Metal Gate Work Function Adjustment for Future CMOS Technology", in Proceedings of the 2001 Symposium on VLSI Technology, Kyoto, Japan, 12–14 June 2001, PP. 45–46.

[32] De, I., D. Johri, A. Srivastava, and C.M. Osburn, "Impact of Gate Work Function on Device Performance at the 50 nm Technology Node", Solid-State Electronics, Vol. 44, PP. 1077–1080, 2000.

[33] Mistry, K., C. Allen, C. Auth, B. Beattie, D. Bergstrom, M. Bost, M. Brazier, M. Buehler, A. Cappellani, R. Chau, et al., "A 45nm Logic Technology with High-k+Metal Gate Transistors, Strained Silicon, 9 Cu Interconnect Layers, 193 nm Dry Patterning, and 100% Pb-free Packaging", in Proceedings of the 2007 IEEE International Electron Devices Meeting, Washington, DC, USA, 10–12 December 2007, PP. 247–250.

[34] Packan, P., S. Akbar, M. Armstrong, D. Bergstrom, M. Brazier, H. Deshpande, K. Dev, G. Ding, T. Ghani, O. Golonzka, et al., "High Performance 32nm Logic Technology Featuring 2nd Generation High-k + Metal Gate Transistors", in Proceedings of the 2009 IEEE International Electron Devices Meeting (IEDM), Baltimore, MD, USA, 7–9 December 2009, PP. 1–4.

[35] Ma, X., H. Yang, W. Wang, H. Yin, H. Zhu, C. Zhao, D. Chen, and T. Ye, "The Effects of Process Condition of Top-TiN and TaN Thickness on the Effective Work Function of MOSCAP with High-k/Metal Gate Stacks", Journal of Semiconductors, Vol. 35, P. 106002, 2014.

[36] Yang, H., W. Luo, L. Zhou, H. Xu, B. Tang, E. Simoen, H. Yin, H. Zhu, C. Zhao, W. Wang, et al., "Impact of ALD TiN Capping Layer on Interface Trap and Channel Hot Carrier Reliability of HKMG nMOSFETs", IEEE Electron Device Letters, Vol. 39, PP. 1129–1132, 2018.

[37] Fenouillet-Beranger, C., S. Denorme, B. Icard, F. Boeuf, J. Coignus, O. Faynot, L. Brevard, C. Buj, C. Soonekindt, J. Todeschini, et al., "Fully-depleted SOI Technology Using High-k and Single-metal Gate for 32 nm Node LSTP Applications Featuring 0.179 μm2 6T-SRAM Bitcell", in Proceedings of the 2007 IEEE International Electron Devices Meeting, Washington, DC, USA, 10–12 December 2007, PP. 267–270.

[38] Deleonibus, S., C. Mazure, P. Gaud, H. Grampeix, J.P. Colonna, B. Previtali, H. Dansas, D. Lafond, C. Jahan, C. Fenouillet-Beranger, et al., "25 nm Short and Narrow Strained FDSOI with TiN/HfO2 Gate Stack", in Proceedings of the 2006 Symposium on VLSI Technology, Honolulu, HI, USA, 13–15 June 2006, PP. 134–135.

[39] Barral, V., T. Poiroux, F. Andrieu, C. Buj-Dufournet, O. Faynot, T. Ernst, L. Brevard, C. Fenouillet-Beranger, D. Lafond, J.M. Hartmann, et al., "Strained FDSOI CMOS Technology Scalability Down to 2.5 nm Film Thickness and 18nm Gate Length with a TiN/HfO2 Gate Stack", in Proceedings of the 2007 IEEE International Electron Devices Meeting, Washington, DC, USA, 10–12 December 2007, PP. 61–64.

[40] Doris, B., Y.H. Kim, B.P. Linder, M. Steen, V. Narayanan, D. Boyd, J. Rubino, L. Chang, J. Sleight, A. Topol, et al., "High Performance FDSOI CMOS Technology with Metal Gate and High-k", in Proceedings of the 2005 Symposium on VLSI Technology, Kyoto, Japan, 14–16 June 2005, PP. 214–215.

[41] Andrieu, F., O. Faynot, X. Garros, D. Lafond, C. Buj-Dufournet, L. Tosti, S. Minoret, V. Vidal, J.C. Barbe, F. Allain, et al., "Comparative Scalability of PVD and CVD TiN on HfO2 as a Metal Gate Stack for FDSOI cMOSFETs Down to 25nm Gate Length and Width", in Proceedings of the 2006 International Electron Devices Meeting, San Francisco, CA, USA, 11–13 December 2006, PP. 1–4.

[42] Collaert, N., M. Demand, I. Ferain, J. Lisoni, R. Singanamalla, P. Zimmerman, Y.S. Yim, T. Schram, G. Mannaert, M. Goodwin, et al., "Tall Triple-gate Devices with TiN/HfO2 Gate Stack", in Proceedings of the Digest of Technical Papers, 2005 Symposium on VLSI Technology, 2005, Kyoto, Japan, 14–16 June 2005, PP. 108–109.

[43] Liu, Y., S. Kijima, E. Sugimata, M. Masahara, K. Endo, T. Matsukawa, K. Ishii, K. Sakamoto, T. Sekigawa, H. Yamauchi, et al., "Investigation of the TiN Gate Electrode with Tunable Work Function and Its Application for FinFET Fabrication", IEEE Transactions on Nanotechnology, Vol. 5, PP. 723–730, 2006.

[44] Matsukawa, T., C. Yasumuro, H. Yamauchi, S. Kanemaru, M. Masahara, K. Endo, E. Suzuki, and J. Itoh, "Work Function Control of Al-Ni Alloy for Metal Gate Application", in Extended Abstracts of the 2004 International Conference on Solid State Devices and Materials, Tokyo, 14–17 September 2004, PP. 464–465.

[45] Han, K., X. Ma, H. Yang, and W. Wang, "Modulation of the Effective Work Function of a TiN Metal Gate for NMOS Requisition with Al Incorporation", Journal of Semiconductors, Vol. 34, P. 076003, 2013.

[46] Kesapragada, S., R. Wang, D. Liu, G. Liu, Z. Xie, Z. Ge, H. Yang, Y. Lei, X. Lu, X. Tang, et al., "High-k/Metal Gate Stacks in Gate First and Replacement Gate Schemes", in Proceedings of the 2010 IEEE/SEMI Advanced Semiconductor

Manufacturing Conference (ASMC), San Francisco, CA, USA, 11–13 July 2010, PP. 256–259.
[47] Skotnicki, T., G. Merckel, and T. Pedron, "The Voltage-Doping Transformation a New Approach to the Modelling of MOSFET Short-Channel Effects", in Proceedings of the 17th European Solid State Device Research Conference, Bologna, Italy, 14–17 September 1987, PP. 543–546.
[48] Skotnicki, T., "Heading for Decananometer CMOS—Is Navigation Among Icebergs Still a Viable Strategy?", in Proceedings of the 30th European Solid-State Device Research Conference, Cork, Ireland, 11–13 September 2000, PP. 19–33.
[49] Colinge, J.-P., "The SOI MOSFET: From Single Gate to Multigate", in FinFETs and Other Multi-Gate Transistors, Colinge, J.-P., Ed.; Springer, Boston, MA, 2008, PP. 1–48. https://doi.org/10.1007/978-0-387-71752-4_1.
[50] Hisamoto, D., L. Wen-Chin, J. Kedzierski, E. Anderson, H. Takeuchi, K. Asano, K. Tsu-Jae, J. Bokor, and H. Chenming, "A Folded-channel MOSFET for Deep-sub-tenth Micron Era", in Proceedings of the International Electron Devices Meeting 1998, San Francisco, CA, USA, 6–9 December 1998, PP. 1032–1034.
[51] Nagy, D., G. Indalecio, A.J. García-Loureiro, M.A. Elmessary, K. Kalna, and N. Seoane, "FinFET Versus Gate-All-Around Nanowire FET: Performance, Scaling, and Variability", IEEE Journal of the Electron Devices Society, Vol. 6, PP. 332–340, 2018.
[52] Natarajan, S., M. Agostinelli, S. Akbar, M. Bost, A. Bowonder, V. Chikarmane, S. Chouksey, A. Dasgupta, K. Fischer, Q. Fu, et al., "A 14nm Logic Technology Featuring 2nd-generation FinFET, Air-gapped Interconnects, Self-aligned Double Patterning and a 0.0588 µm2 SRAM Cell Size", in Proceedings of the 2014 IEEE International Electron Devices Meeting, San Francisco, CA, USA, 15–17 December 2014, PP. 3.7.1–3.7.3.
[53] Mertens, H., R. Ritzenthaler, A. Hikavyy, M.S. Kim, Z. Tao, K. Wostyn, S.A. Chew, A.D. Keersgieter, G. Mannaert, E. Rosseel, et al., "Gate-all-around MOSFETs Based on Vertically Stacked Horizontal Si Nanowires in a Replacement Metal Gate Process on Bulk Si Substrates", in Proceedings of the 2016 IEEE Symposium on VLSI Technology, Honolulu, HI, USA, 14–16 June 2016, PP. 1–2.
[54] Mertens, H., R. Ritzenthaler, A. Chasin, T. Schram, E. Kunnen, A. Hikavyy, L. Ragnarsson, H. Dekkers, T. Hopf, K. Wostyn, et al., "Vertically Stacked Gate-all-Around Si Nanowire CMOS Transistors with Dual Work Function Metal Gates", in Proceedings of the 2016 IEEE International Electron Devices Meeting (IEDM), San Francisco, CA, USA, 3–7 December 2016, PP. 19.17.11–19.17.14.
[55] Singh, S., and B. Raj, "Analytical Modeling and Simulation Analysis of T-shaped III-V Heterojunction Vertical T-FET", Superlattices and Microstructures, Elsevier, Vol. 147, PP. 106717, November 2020.
[56] Chawla, T., M. Khosla, and B. Raj, "Optimization of Double-gate Dual Material GeOI-Vertical TFET for VLSI Circuit Design", IEEE VLSI Circuits and Systems Letter, Vol. 6, No. 2, PP. 13–25, August 2020.
[57] Kaur, M., N. Gupta, S. Kumar, B. Raj, and Arun Kumar Singh, "RF Performance Analysis of Intercalated Graphene Nanoribbon Based Global Level Interconnects", Journal of Computational Electronics, Springer, Vol. 19, PP. 1002–1013, June 2020.

[58] Wadhwa, G., and B. Raj, "An Analytical Modeling of Charge Plasma Based Tunnel Field Effect Transistor with Impacts of Gate Underlap Region", Superlattices and Microstructures, Elsevier, Vol. 142, PP. 106512, June 2020.
[59] Singh, S., and B. Raj, "Modeling and Simulation Analysis of SiGe Hetrojunction Double Gate Vertical T-shaped Tunnel FET", Superlattices and Microstructures, Elsevier, Vol. 142, PP. 106496, June 2020.
[60] Singh, S., and B. Raj, "A 2-D Analytical Surface Potential and Drain Current Modeling of Double-Gate Vertical T-shaped Tunnel FET", Journal of Computational Electronics, Springer, Vol. 19, PP. 1154–1163, April 2020.
[61] Singh, S., S. Bala, B. Raj, and B. Raj, "Improved Sensitivity of Dielectric Modulated Junctionless Transistor for Nanoscale Biosensor Design", Sensor Letter, ASP, Vol. 18, PP. 328–333, April 2020.
[62] Kumar, V., S. Kumar, and B. Raj, "Design and Performance Analysis of ASIC for IoT Applications", Sensor Letter, ASP, Vol. 18, PP. 31–38, January 2020.
[63] Wadhwa, G., and B. Raj, "Design and Performance Analysis of Junctionless TFET Biosensor for High Sensitivity", IEEE Nanotechnology, Vol. 18, PP. 567–574, 2019.
[64] Wadhera, T., D. Kakkar, G. Wadhwa, and B. Raj, "Recent Advances and Progress in Development of the Field Effect Transistor Biosensor: A Review", Journal of Electronic Materials, Springer, Vol. 48, No. 12, PP. 7635–7646, December 2019.
[65] Singh, S., and B. Raj, "Design and Analysis of Hetrojunction Vertical T-shaped Tunnel Field Effect Transistor", Journal of Electronics Material, Springer, Vol. 48, No. 10, PP. 6253–6260, October 2019.
[66] Goyal, C., J.S. Ubhi, and B. Raj, "A Low Leakage CNTFET Based Inexact Full Adder for Low Power Image Processing Applications", International Journal of Circuit Theory and Applications, Wiley, Vol. 47, No. 9, PP. 1446–1458, September 2019.
[67] Sharma, S.K., B. Raj, and M. Khosla, "Enhanced Photosensivity of Highly Spectrum Selective Cylindrical Gate In1-xGaxAs Nanowire MOSFET Photodetector", Modern Physics Letter-B, Vol. 33, No. 12, PP. 1950144, 2019.
[68] Singh, J., and B. Raj, "Design and Investigation of 7T2M NVSARM with Enhanced Stability and Temperature Impact on Store/Restore Energy", IEEE Transactions on Very Large Scale Integration Systems, Vol. 27, No. 6, PP. 1322–1328, June 2019.
[69] Bhardwaj, A.K., S. Gupta, B. Raj, and Amandeep Singh, "Impact of Double Gate Geometry on the Performance of Carbon Nanotube Field Effect Transistor Structures for Low Power Digital Design", Computational and Theoretical Nanoscience, ASP, Vol. 16, PP. 1813–1820, 2019.
[70] Goyal, C., J. Subhi, and B. Raj, "Low Leakage Zero Ground Noise Nanoscale Full Adder Using Source Biasing Technique", Journal of Nanoelectronics and Optoelectronics, American Scientific Publishers, Vol. 14, PP. 360–370, March 2019.
[71] Singh, A., M. Khosla, and B. Raj, "Design and Analysis of Dynamically Configurable Electrostatic Doped Carbon Nanotube Tunnel FET", Microelectronics Journal, Elesvier, Vol. 85, PP. 17–24, March 2019.
[72] Goyal, C., J.S. Ubhi, and B. Raj, "A Reliable Leakage Reduction Technique for Approximate Full Adder with Reduced Ground Bounce Noise", Journal of

Mathematical Problems in Engineering, Hindawi, Vol. 2018, PP. 16, Article ID 3501041, 15 October 2018.
[73] Wadhwa, G., and B. Raj, "Label Free Detection of Biomolecules using Charge-Plasma-Based Gate Underlap Dielectric Modulated Junctionless TFET", Journal of Electronic Materials (JEMS), Springer, Vol. 47, No. 8, PP. 4683–4693, August 2018.
[74] Wadhwa, G., and B. Raj, "Parametric Variation Analysis of Charge-Plasma-based Dielectric Modulated JLTFET for Biosensor Application", IEEE Sensor Journal, Vol. 18, No. 15, 1 August 2018.
[75] Yadav, D., S.S. Chouhan, S.K. Vishvakarma, and B. Raj, "Application Specific Microcontroller Design for IoT based WSN", Sensor Letter, ASP, Vol. 16, PP. 374–385, May 2018.
[76] Singh, G., R.K. Sarin, and B. Raj, "Fault-Tolerant Design and Analysis of Quantum-Dot Cellular Automata Based Circuits", IEEE/IET Circuits, Devices & Systems, Vol. 12, PP. 638–64, 2018.
[77] Singh, J., and B. Raj, "Modeling of Mean Barrier Height Levying Various Image Forces of Metal Insulator Metal Structure to Enhance the Performance of Conductive Filament Based Memristor Model", IEEE Nanotechnology, Vol. 17, No. 2, PP. 268–267, March 2018 (SCI).
[78] Jain, A., S. Sharma, and B. Raj, "Analysis of Triple Metal Surrounding Gate (TM-SG) III-V Nanowire MOSFET for Photosensing Application", Opto-Electronics Journal, Elsevier, Vol. 26, No. 2, PP. 141–148, May 2018.
[79] Jain, N., and B. Raj, "Parasitic Capacitance and Resistance Model Development and Optimization of Raised Source/Drain SOI FinFET Structure for Analog Circuit Applications", Journal of Nanoelectronics and Optoelectronins, ASP, USA, Vol. 13, PP. 531–539, April 2018.

Chapter 4

The Transition from MOSFET to MBCFET

Fabrication and Transfer Characteristics

Amarah Zahra, Ashish Raman, Shamshad Alam, and Balwinder Raj

CONTENTS

4.1 INTRODUCTION

A MOSFET, also known as a metal–oxide–silicon field-effect transistor or metal–oxide–semiconductor field-effect transistor or an insulated gate field-effect transistor (IGFET), is a three-terminal device. A field-effect transistor (FET) can be operated in both depletion and enhancement modes [1].

Jordon Moore predicted in 1965 that after every 18 months, the number of transistors per chip increases as shown in Figure 4.1, and with this increase in number of transistors, the transistor size decreases. Moore's law is not actually a law as it is an observation that has become a self-fulfilling prophecy in the semiconductor industry. Planer transistors have been used for generations with voltage scaling to save the power and reduce

DOI: 10.1201/9781003311379-4

Figure 4.1 Practical MOSFET.

the operating voltage. The process of reducing the horizontal and vertical dimensions of MOSFETs is called scaling. In order to satisfy Moore's law, the channel length (L) and width (W) of the MOSFET have to be lessened by a factor of 0.7%. When the length and width of the MOSFET are reduced by 0.7%, the area of the MOSFET that is L×W is reduced by half. MOSFET scaling comes with many advantages, such as increased component density, reduction in power consumption, increased speed, and cost per chip decreases. But it has several disadvantages too, such as scaling it down too much can cause problems like second-order effects also called short channel effects. The following are the short channel effects that we encounter in MOSFETs:

1. Drain-induced barrier lowering (DIBL)
2. Threshold voltage roll-off
3. Hot carrier effect
4. Punchthrough effect
5. Gate tunneling current
6. Subthreshold leakage current

So scaling it down was no longer a solution as industrialists had to face many limitations. Fin field-effect transistors (FinFETs) were the solution and it made further voltage scaling possible, and it provided better performance than planer FETs because of their remarkably improved electrostatics [2]. FinFETs feature a wider range of design possibilities and greater channel management, allowing for lower power designs and improved noise tolerance [3]. FinFET devices are used to decrease leakage current and power dissipation as the front and back gates can be controlled separately or both simultaneously, and thus, it can enhance the performance of the FinFET technology [2]. In FinFETs, the current is controlled by applying a gate on each of the three sides of a fin. Each fin has a definite shape, size, and width.

Scaling FinFET decreases sideway dimensions to enlarge device density per unit area while enlarging fin height to upgrade device performance. A chip is made up of three main components: a transistor, contacts, and interconnects. The interconnects, which are located on top of the transistor, are made up of small copper wiring schemes that transport electrical impulses from one transistor to the next. A layer termed the middle-of-line (MOL) connects the transistor and the interconnect. The MOL is made up of teeny-tiny contact structures. In order to achieve ideal electrostatics, the channel must be completely covered by a gate. However, industrialists were again confronted with difficulties because lowering the voltage below 0.75V with this architecture has proved extremely difficult. Also, backend-of-the-line (BEOL) and MOL are the constraints in advanced chips. In the MOL, contact resistance is a problem. Copper interconnects for semiconductors are manufactured in the BEOL. At each node, the interconnects are becoming more compact, resulting in resistance-capacitance (RC) delays in the chips. RC delay will be an issue here. There are other obstacles. When the fin width hits 5nm, FinFETs run out of steam. FinFETs with dimensions of 5nm and 3nm are approaching these limits. Despite the fact that the gate controls three sides of the fin, one side remains uncontrolled. As the gate length is shortened, more short channel effects and leakage through the device's uncontested bottom ensue, and smaller devices are unable to fulfill power and performance targets. This thing forced the industrialist and researchers to implement and innovate this new technology called GAA technology. As FinFETs need a supplementary area to improve the speed. Fins are to be added laterally in the case of FinFET technology, which undoubtedly improves speed but at the expense of area.

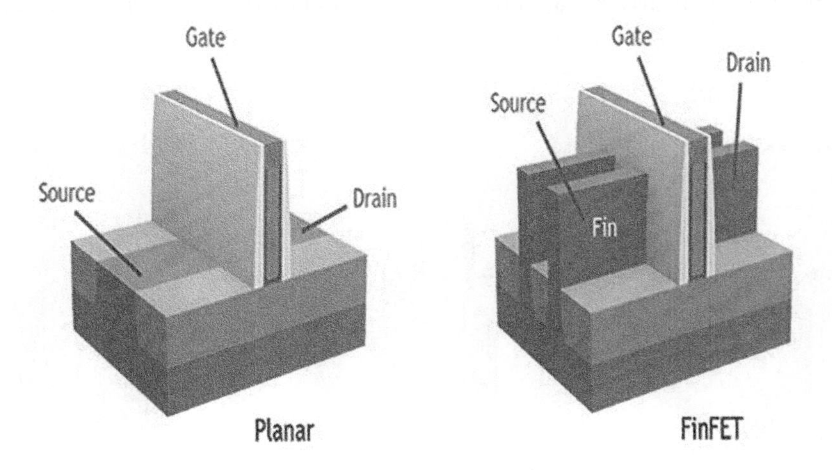

Figure 4.2 The structure of planer FET and FinFET technology.

Taking these constraints; roadblocks like speed, area, and power consumption, and reducing operating voltage into consideration, companies came up with GAA and then shifted to MBCFET technology.

4.1.1 GAA

GAA transistors are an altered transistor structure in which the channel is fully encased by a gate. This GAAFET technology is providing the best solutions to the problems faced in planer FET, such as the second-order effect and the increase in area for FinFET technology [4]. FinFET technology is employed at the 14nm, 10nm, and 7nm nodes. Vertical "fins" are positioned above the previous two-dimensional (2D) channel structure, increasing the contact area between the transistor channel and the gate, whereas in GAA, either nanowires or nanosheets are used. But various problems are associated with nanowires. Nanowires are hard to fabricate yet ideal for low-power applications. They require very high temperatures for their manufacture (600–900 degrees Celsius); only anodes are manufactured from nanowires, and they are very expensive. So nanosheets have many advantages over nanowires in terms of performance and scaling. The GAA is made up of separate horizontal sheets encircled on all sides by gate materials. In comparison to FinFETs, this gives better channel control. Unlike FinFETs, which require numerous side-by-side fins to carry more current, GAA transistors can carry more current by stacking a few nanosheets vertically and wrapping gate material across the channels.

The dimensions of nanosheets may be scaled, allowing transistors to be designed to meet specific performance requirements. The performance is improved by 35%, the power consumption is reduced by 50%, the area is reduced by nearly 45%, and the benefits don't stop there. The operating

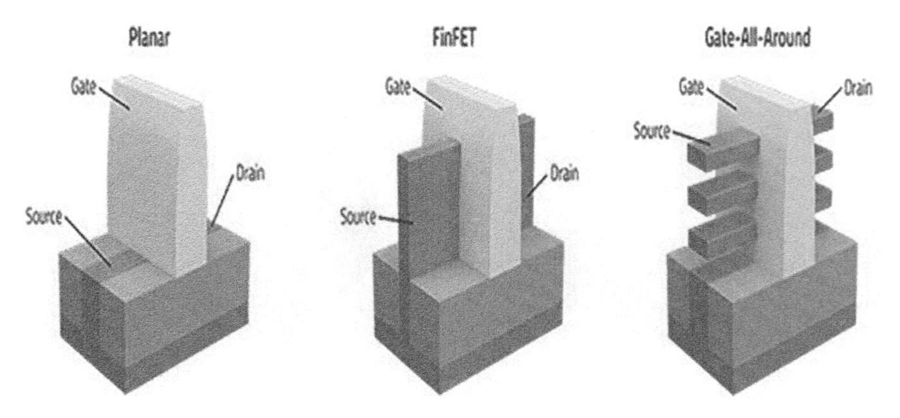

Figure 4.3 Structure difference between planer FET, FinFET, and GAA technology.

voltage is reduced to 0.1V in the 3nm technology that uses nanosheets. So the other variant of GAA technology is the multibridge-channel field-effect transistor, or MBCFET.

4.1.2 Nanosheets

A nanosheet is a three-dimensional nanostructure that has a thickness of 1 to 100nm. A nanosheet is made up of numerous tiny horizontal parts or sheets that are stacked vertically. A canal is formed by each sheet [5]. Because the current is controlled on four sides of the structure, nanosheet FETs give better performance with less leakage. FinFETs can only be quantized with a limited number of fins, posing certain design challenges. A nanosheet has the advantage of being able to have varied nanosheet widths. The drive current of a transistor with a wider sheet is higher. With a narrow sheet, a smaller device can be made with less drive current. A nanosheet and a nanowire are linked. The channels are made up of wires rather than sheets. The channel width is restricted, resulting in a lower drive current. To avoid lattice deformation and many other faults, the germanium concentration of the SiGe layers should be as low as feasible.

However, because etch selectivity rises with Ge concentration, degradation of the silicon layers undergoing the internal spacer compression or channel release carve will impact the channel width and thus threshold voltage [6]. In contrast to vertical GAA NW and NS, horizontal GAA nanowire and nanosheet can be produced with minimal divergence from FinFET devices [7]. Although GAAFETs have greater potential for profound scaling than FinFETs, Omega-FETs, dual-gate MOSFETs, and single-gate MOSFETs, one of their primary downsides is its weak subthreshold characteristic, which restricts their application in reduced power and steep switching

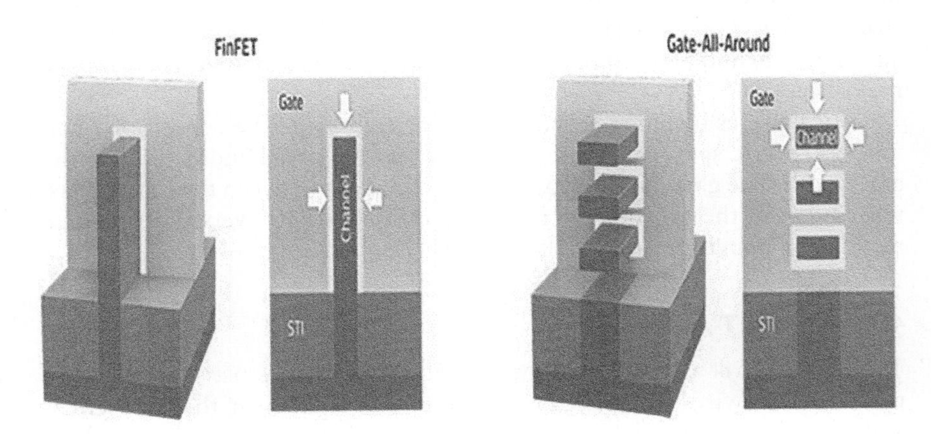

Figure 4.4 Internal structure of FinFET and GAA technology.

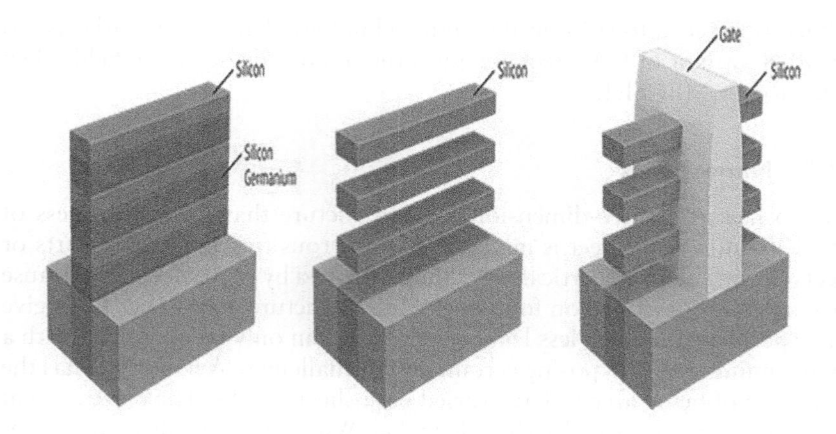

Figure 4.5 GAA FET.

circuits [8]. Short channel effects are not a problem with GAA-FET, and it can be used to replace the fin structure in future nodes. The cylindrical channel GAAFET's ON-to-OFF current ratio is about five times better than the square channel GAAFET's. The channel's cylindrical shape can increase a GAAFET's performance even further. In terms of performance and power efficiency, the GAA structure is perfect for portable devices [9].

4.1.3 Manufacturing Challenges

Nanosheets are simple in concept, but they pose novel production obstacles. Fabricating the structure is one of the difficulties. The major construction difficulties occur as a result of the complicated structure that is being constructed. The GAA transistors are made by developing a super lattice of alternating Si and SiGe epitaxial layers that serves as the foundation for the nanosheets.

Critical stages include the channel release etches and the deposition of an inner dielectric spacer to protect the source/drain areas and determine the gate width. The gate dielectric and metal, including between the nanosheets, must next fill the area left by removing the sacrificial layers. For the gate metal, new materials are likely to be used.

4.2 MBCFET

The MBCFET is a special variant of GAA in which we use nanosheets in place of nanowires. An MBCFET is more power-efficient than the GAA, and its performance is subsequently better. MBCFETs are formed of nanosheets, thus allowing large current to flow through it while having a

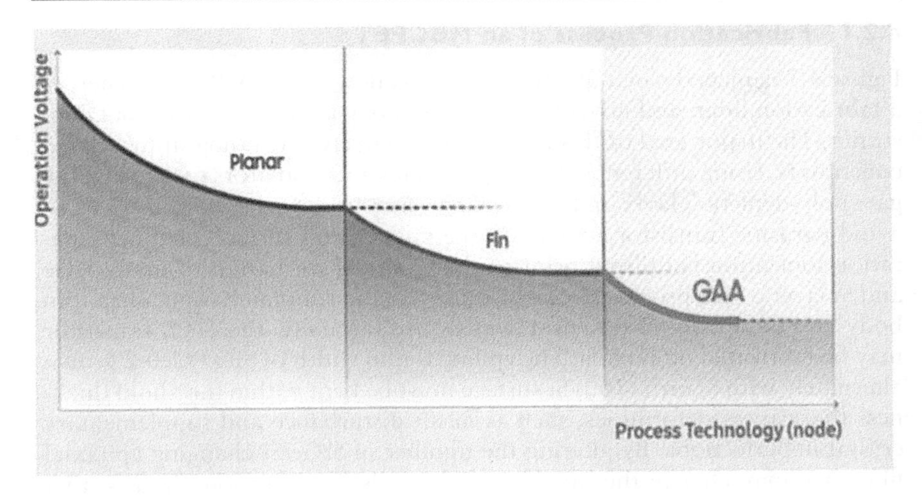

Figure 4.6 Reduction in operating voltage with the advent in technology.

simpler device integration. Depending on whether we want high performance or performance, it can be obtained by modulating the device channel width. MBCFETs are compatible with a FinFET design, and designers can easily replace FinFETs with MBCFETs. It has several advantages, such as improved epitaxial growth, increased strain, and reduced resistance, leading to more current. Here, the operating voltage is reduced to 0.1V. Hence, lower operating voltages help use power in an efficient manner.

From Figure 4.6, it is clear that as we scale down and as we shift from a planar MOSFET to GAA, our operating voltage is reduced. Hence, our performance will get better.

A MBCFET provides greater advantages than a traditional MOSFET. The current drivability of an MBCFET is 4.6 times that of a plane MOSFET. Because of the narrow bulk enclosed by the gate, the subthreshold swing (SS) of the MBCFET is 61 mV/dec, which is virtually optimal [10]. With only a few changed masks, MBCFETs may be produced utilizing 90% or more of FinFET methods, allowing for a simple migration from FinFETs. MBCFETs' performance is much better with a 65-mV/dec SS at a small gate length, greater DC efficiency with a bigger operative channel width (Weff) in a standard layout, and pattern freedom with different nanosheet widths [11]. MBCFETs have various advantages, including increased physical scalability owing to the damascene gate technique, superior short channel effect immunity owing to the GAA structure, and significant current drivability due to the enlarged vertical channel width and transportability. Because a bulk substrate is used, the cost per functionality is low, and the quality is good. Due to the self-aligned double-gate technique, manufacturability is improved [12].

4.2.1 Fabrication Process of an MBCFET

Figure 4.7 depicts the manufacturing process flow for an MBCFET. There is a fabrication limit, and novel process technology is one answer to that constraint. The major goal of this approach is to overcome lithography-related constraints. Using different types of gate plates and insulators to decrease the gate poly-depletion layer and gate leakage current are other options [13]. To avoid parasitic transistor activity on the bulk silicon surface, channel separation ionization is utilized prior to the epitaxial formation of many SiGe and Si layers. This procedure can be removed and simplified using ultra-thin body (UTB) silicon-on-insulator wafers. Additionally, the UTB transistor may be fabricated on bulk Si. The epitaxial film width of SiO.8Ge0.2 grown alternately with Si on a (100) Si surface must be kept within threshold thickness that causes deformities, such as misfit disturbance and supplementary crystal imperfections. By altering the number of SiGe/Si-changing epitaxial films, we may change the number of channels, or inversion charges, that impact MOSFET's current drivability [10].

An oxide bogus gate is used as a tough mask to carve the epitaxial layers in the source/drain region. After carving the source/drain area, Si selective

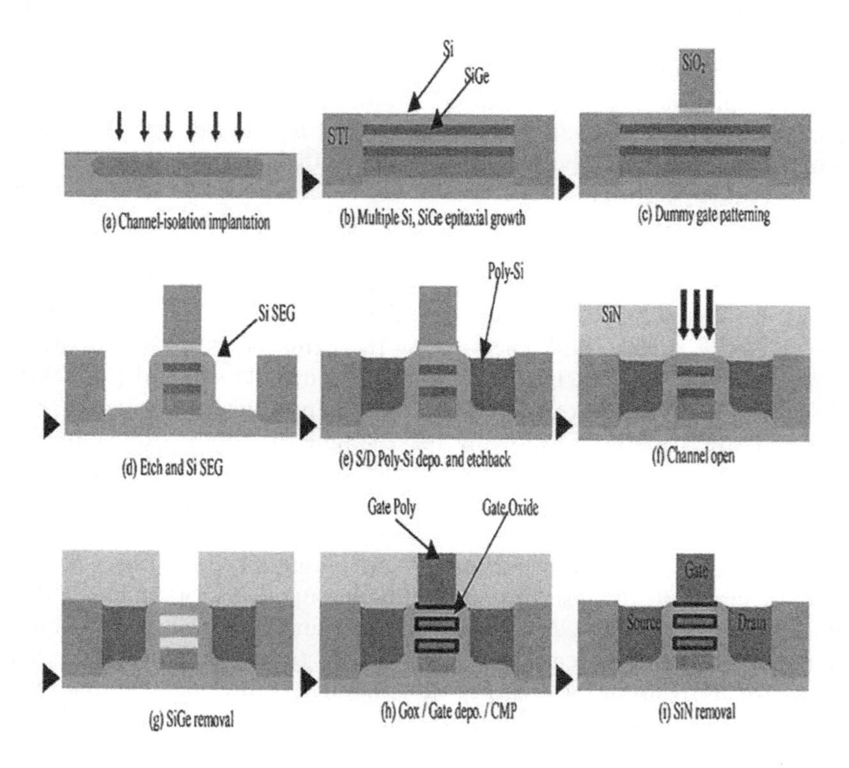

Figure 4.7 Fabrication process of MBCFET.

epitaxial growth (SEG) is used to generate a thin source–drain extension (SDE) layer, as illustrated in Figures 4.7 and 4.8. Tilted ion implantation is used to dope the SDE layer. After poly-Si coating to cover the source/drain region, chemical mechanical polishing (CMP) and etch back of poly-Si are conducted to planarize the source/drain poly-Si, as shown in Figure 4.7. Then, using CMP, dense SiN is coated and planarized. The multibridge channel area is injected with numerous energies to change the threshold voltage once the dummy gate is removed. SiN and Si are used as hard masks to anisotopically etch the oxide of shallow trench isolation, followed by selective Sio.8 Geo.2 removal, to construct the structure displayed in Figure 4.7. After this operation is complete, Figure 4.9 depicts a bird's-eye perspective.

Highly selective Si0.8 Geo.2 removal over Si was one of the fundamental technologies used to make this MBCFET. The Si Ge layers were selectively

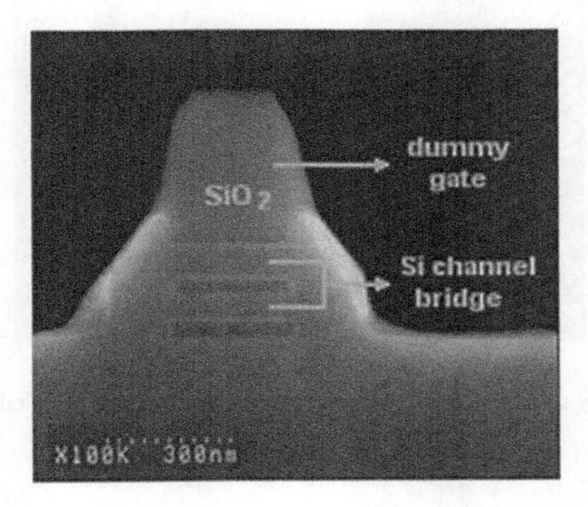

Figure 4.8 Searching electron microscope image after Si SEG of Figure 4.7d.

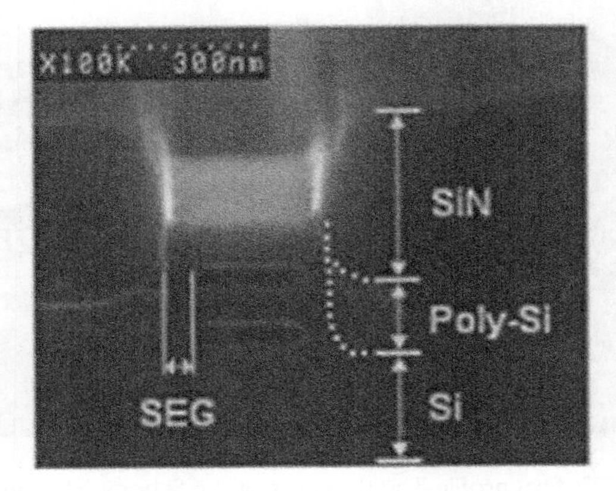

Figure 4.9 Bird's-eye view of Figure 4.7d.

eliminated against Si, as illustrated in Figure 4.10, with a selectivity of greater than 300: 1 [14].

Oxidation of the gate to produce the gate of an n-channel MBCFET, 2.5nm-thick layers of N_2O, N + doped poly-Si deposition, chemical mechanical planarization, and SiN sheets are sequentially applied. Following that, traditional complementary metal–oxide semiconductor (CMOS) methods are used to complete the transistor's manufacturing. Figure 4.11 depicts the

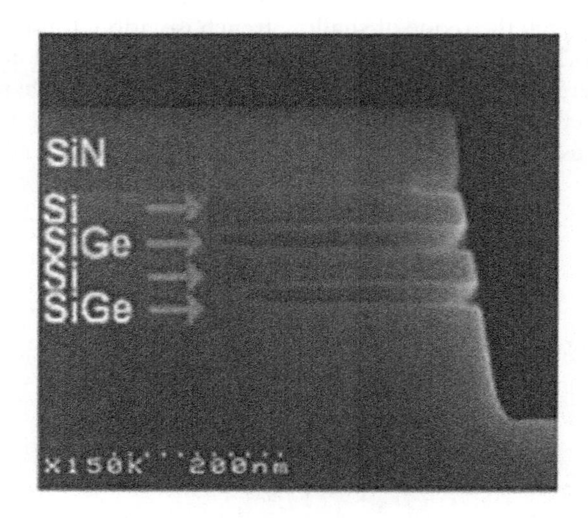

Figure 4.10 A very selective SiGe over Si etching profile has been profitably obtained.

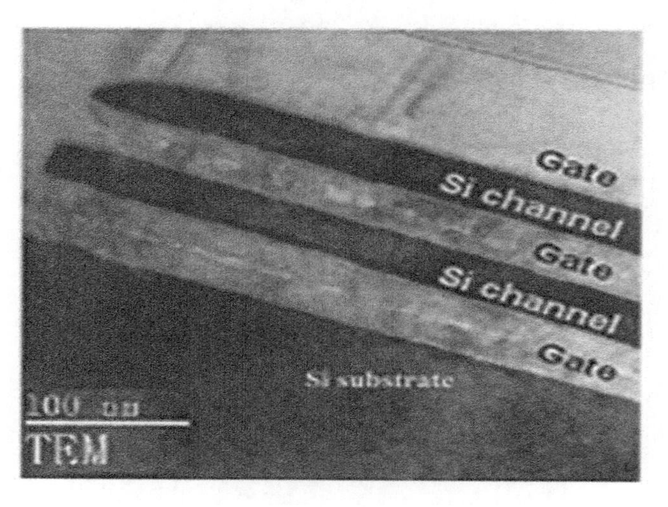

Figure 4.11 MBCFET's final profile. The thickness of the Si body (channel) is 32 nm.

MBCFET's final profile. The thickness of the drifting Si substrate is 32nm. Except for channel isolation and SDE implantation, the insertion methods of a plane MOSFET are identical to those of an MBCFET.

4.2.2 Comparison of Output and Transfer Characteristics of a MOSFET and an MBCFET

The MOS transistor has four terminals, which are referred to as the gate (G), drain (D), source (S), and body (B). The inversion layers in the channel area joins the source and drain junctions. The channel length (L) is determined as the interval connecting the source and drain. Channel width (W) refers to the breadth of a channel in a direction normal to its length [15].

4.2.2.1 *Output Characteristics of MOSFET*

The output characteristics for NMOS and PMOS are demonstrated in Figure 4.13a and Figure 4.13b, respectively. The cutoff zone refers to the area of output characteristics where no current flows. When the channel develops in an NMOS or PMOS transistor, a positive (negative) drain voltage with reference to the source provides a parallel electric field that drives the electrons (holes) in the direction of drain, resulting in a positive (negative) drain current in the transistor.

For electron and hole currents, the positive current convention is employed, but e-are the real charge carriers in both situations. If the

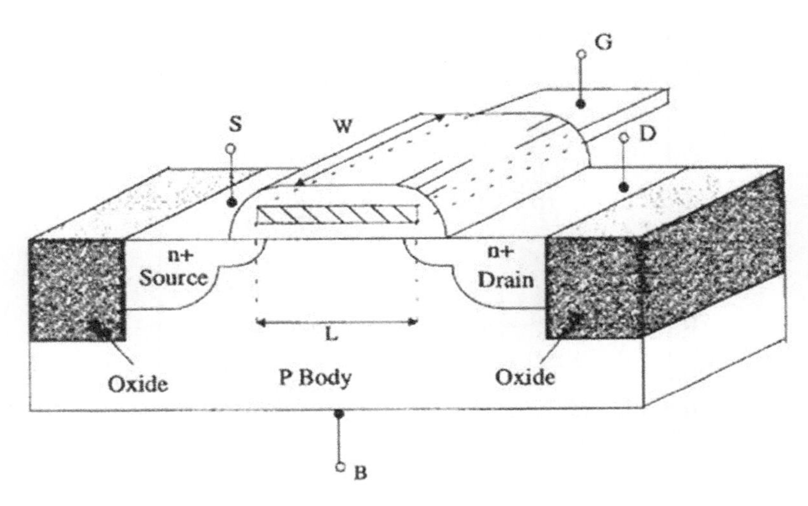

Figure 4.12 Structure of a MOSFET.

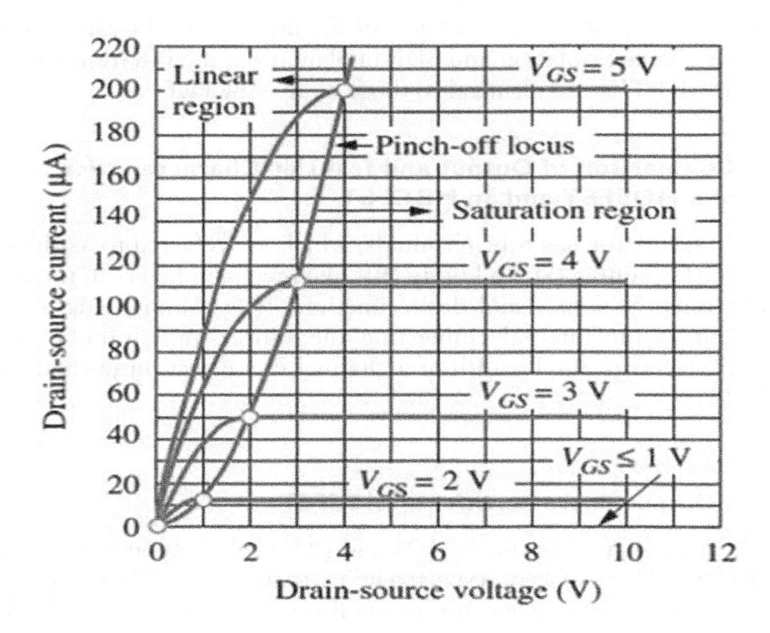

Figure 4.13a Output characteristics for an NMOS transistor.

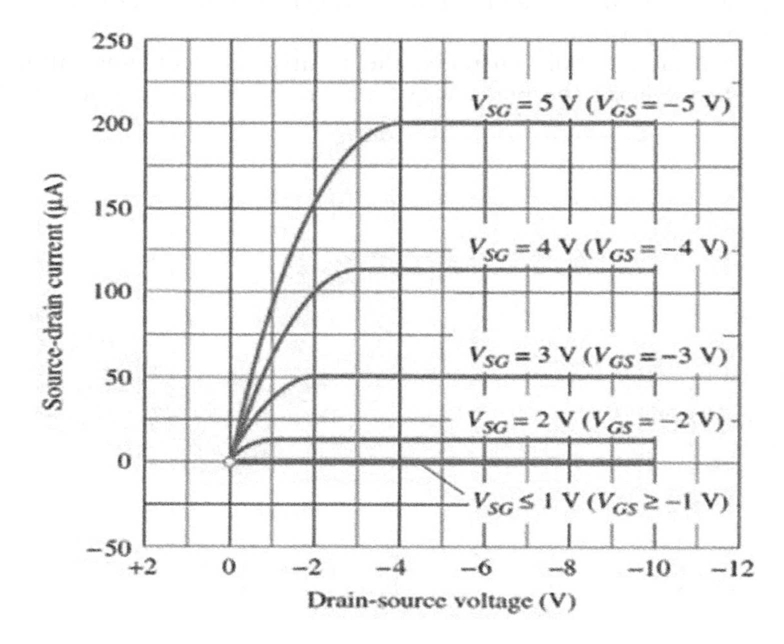

Figure 4.13b Output characteristics for PMOS transistor.

channel horizontal electric field is of the same order or smaller than the vertical thin oxide field, the inversion channel stays nearly uniform over the device distance. This uninterrupted transport pattern from source to drain places the transistor in a bias state that can be referred to as non-saturated, linear, or ohmic. Both the drain and the source are successfully closed off. When VGS > VDS + Vtn for NMOS transistors and VGS < VDS +Vtp for PMOS transistors, this occurs. In the linear bias condition, drain current is linearly proportional to drain-source voltage over small intervals. However, if the NMOS drain voltage exceeds the limit, such that VGS < VDS + Vtn, the horizontal electric field at the drain end becomes greater than the vertical field, causing an imbalance in the channel carrier inversion pattern.

If the drain voltage increases while the gate voltage stays constant, VGD in the drain region might fall below the threshold value. The reversed section of the channel detaches from the drain and no more "touches" this terminal since there is no carrier inversion at the drain-gate oxide area. The channel's pinched-off section generates a depletion zone with a strong electric field. A pn junction is formed by the n-drain and the p-bulk. Whenever this occurs, the inversion layer is deemed "pinched off," and the device is said to be saturated [16].

4.2.2.2 *Transfer Characteristics of a MOSFET*

The transfer characteristic links the drain current (ID) response to the input gate-source-controlling voltage (VGS). The gate current is almost nonexistent since the gate terminal is electrically insulated from the other terminals (drain, source, bulk). As a result, gate current is not included in device characteristics. The gate voltage during which the transistor transmits current and exits the OFF state may be found using the transfer characteristic curve. The device's threshold voltage is this (Vtn). The observed input characteristics of NMOS and PMOS transistors with a low 0.1V voltage between their drain and source terminals are shown in Figure 4.14. The transistors are in their non-saturated bias states. The threshold voltage for the NMOS transistor in Figure 4.14a is achieved when VGS increases, causing ID to rise. ID is almost negative for VGS between 0 and 0.7V, suggesting a very high identical opposition between the drain and source ports.

The current increases fast with VGS after it hits 0.7V, showing that the identical opposition at the drain reduces as the VGS rises. As a result, Vtn 0.7V is the threshold voltage of the supplied NMOS transistor. The PMOS transistor's input characteristic is identical to that of the NMOS transistor in Figure 4.14b, with the exception that the ID and VGS polarities are inverted.

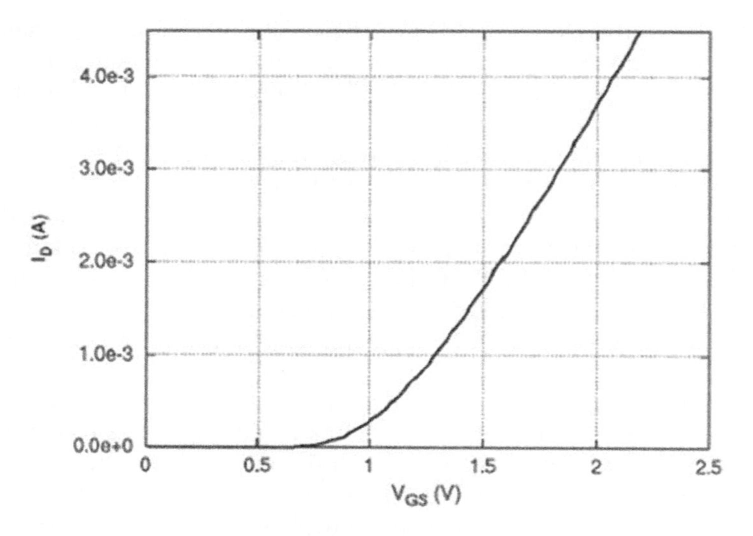

Figure 4.14a Transfer characteristics for an NMOS transistor.

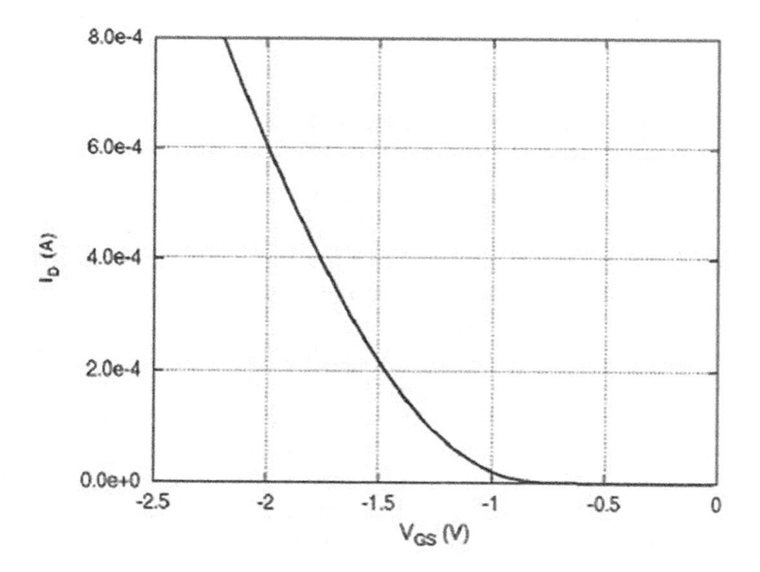

Figure 4.14b Transfer characteristics for PMOS transistor.

4.2.2.3 *Output and Transfer Characteristics of an MBCFET*

The properties of the n-channel MBCFET are compared to those of the n-channel plane MOSFET in Figures 4.15 and 4.16. The Vth of a plane MOSFET is 0.45V, whereas the Vth of an MBCFET is 0.1V. At VGS = Vds = 1V,

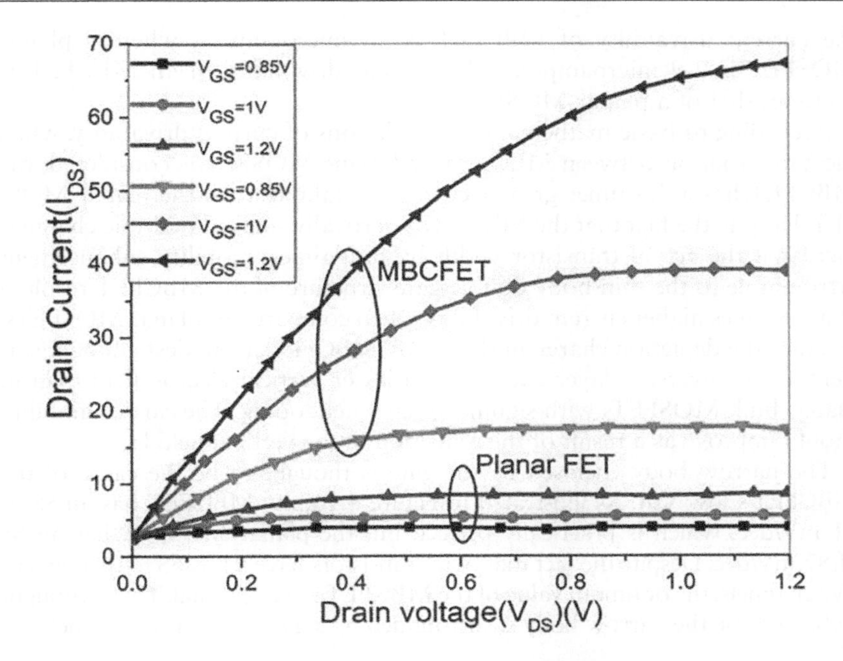

Figure 4.15 Output characteristics for an MBCFET and a MOS transistor.

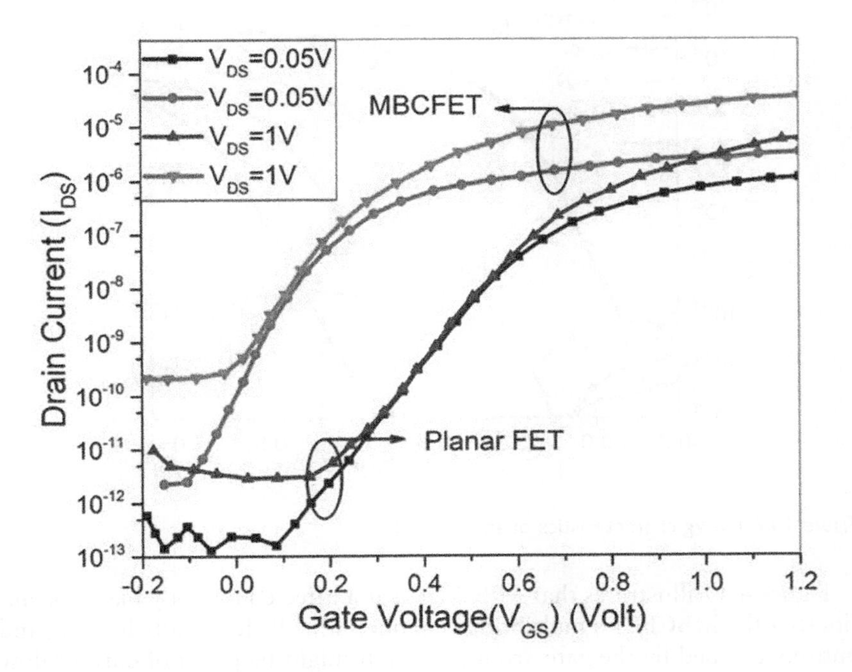

Figure 4.16 ID–VG characteristics of an MBCFET.

the current drivability of MBCFET is 38 microampere, whereas planar MOSFET is 2.9 microampere. The current drivability of an MBCFET is 13 times that of a planar MOSFET.

According to basic mathematical calculations of current drivability, when the Vth variation between MBCFET and plane MOSFET is considered, the MBCFET has a 4.6 times greater current drivability than the planar MOSFET. Despite the fact that the MBCFET's vertically stacked multiple channels increased the actual transistor width by four times, mobility enhancement attributable to the thin-body double-gate structure of the MBCFET resulted in a 4.6 times higher current drivability when compared to planar MOSFETs. Because the depletion charge in thin-body MBCFETs is modest enough, carriers in the inversion layer experience a lesser vertical electric field than in planar bulk MOSFETs with significant channel doping. The carrier mobility should improve as a result of the reduction in the vertical field [17].

The narrow body enclosed by the gate is thought to be the cause of the MBCFET's low Vth. As illustrated in Figure 4.16, the MBCFET has an SS of 61 mV/dec, which is practically perfect, but the planar MOSFET has an SS of 87 mV/dec. Despite the fact that both transistors have a low SS to their extensive channels, the optimum value of the MBCFET is exceptional. This is thought to be because the narrow body is surrounded by a gate and gentle channel.

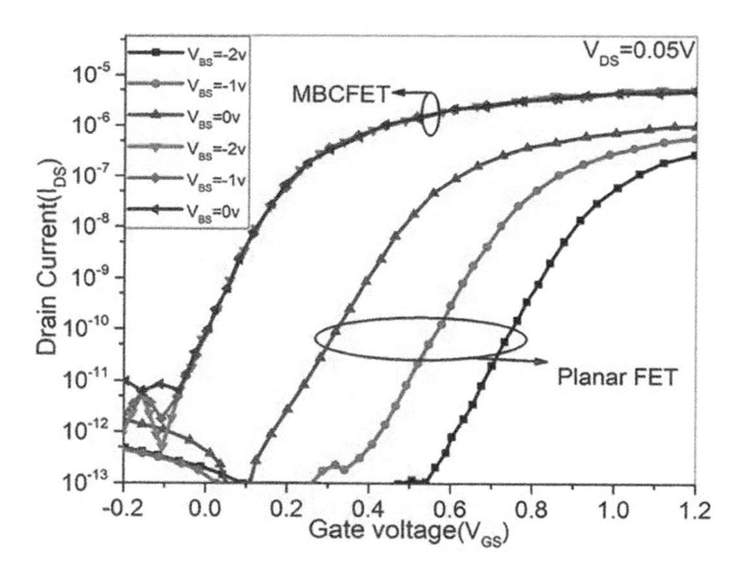

Figure 4.17 ID–Vg characteristics of an MBCFET.

Figure 4.17 illustrates that MBCFET is not affected by body bias. It signifies that the MBCFET's multibridge channels are all electrically floating and entirely drained by the gate around them. It might be proof of current flow in the MBCFET's floating channels rather than the body [10].

4.2.3 Leakage Profile Improvement of an MBCFET

The leakage profile of a typical MBCFET can be enhanced by adding a core insulator layer. The fluctuation in electron current density is depicted in Figures 4.18a and 4.18b, which show the cross-sectional view of electron current density for conventional and core insulator MBCFETs, respectively. The channel of a core insulator MBCFET has a narrow insulator (oxide) film in the center, and the channel is bordered by HfO2 gate oxide and a metal gate [18].

Hence, because of the cross stacking of Si and Si0.8Ge0.2 layers, as well as an insulator layer between Si and Si, the distance between the gate electrode and the channel increases, reducing the effect of gate biasing on the channel and resulting in poor gate controllability and leakage characteristics, as shown in Figure 4.18a; that is, as the distance from the gate increases, the electrification of the channel decreases. Figure 4.18 shows how to remedy this problem by eliminating the bad gate control region with a core insulator, Figure 4.18b. The MBCFET's electrostatic characteristics are enhanced by adding the core insulator.

4.3 CONCLUSION

This chapter deals with the fabrication process of MBCFETs using a conventional CMOS process. Because of the vertically stacked double-bridge channels, the MBCFET can drive 4.6 times more current than a normal plane MOSFET at about the same threshold voltage. Through the implementation

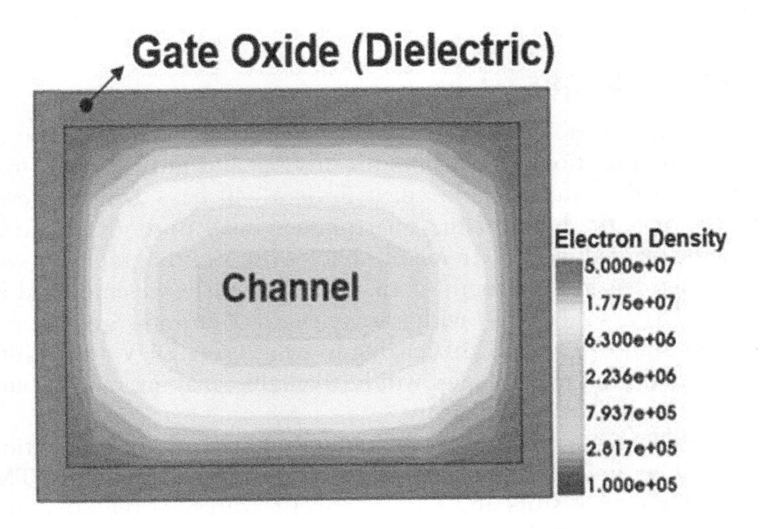

Figure 4.18a Channel cross-sectional view of electron current density of a conventional MBCFET.

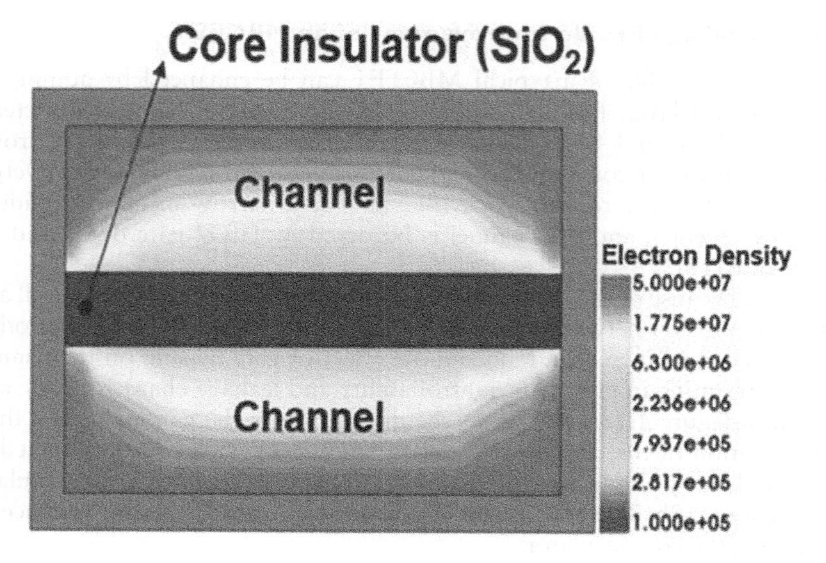

Figure 4.18b Channel cross-sectional view of electron current density of a core insulator MBCFET.

of a simple oxidation procedure, the leakage and electrostatic properties were enhanced. The most appealing feature is that it can improve device performance by a simple oxidation process rather than modifying a complicated photo-patterning process. The suggested approach may greatly reduce the driving current through the core insulator while suppressing leakage current.

4.4 FUTURE SCOPE

MBCFETs will create a boom in the world of artificial intelligence (AI), wireless communication, in autonomous driving, and in high-performance computing. In the future, Apple's next devices will have 2nm processors. The processor is the brain of the smartphone. Small portions of the CPU may now be printed using extreme ultraviolet (EUV) lithography. It works with light that has a wavelength of 13.5nm. Previously, we employed deep ultraviolet (DUV) lithography with a wavelength of 193nm. Small portions of 7nm and 5nm CPUs were challenging to print using DUV. In the future, 3nm, 2nm, and 1nm technologies will be available, making mobile phones more powerful and energy-efficient.

TMSC, a silicon manufacturing company, estimates volume production of 3nm technology in 2022. To accomplish complete node scalability, TMSC will use FinFET transistors and rely on novel features. At the same power levels, performance will improve by 10–15% with 3nm technology, while power consumption will be reduced by 25–30% at the same transistor speeds. In the future, a 1.7× gain in logic area density is predicted, implying a 0.58× scaling ratio between 5nm and 3nm logic [19]. Nvidia has improved the automated arrangement of its standard cells at advanced process nodes at 5nm and 3nm using reinforcement learning and other forms of AI. Samsung is currently working on 3nm technology using MBCFETs, whereas Intel is working on super-fin technology to make a 3nm processor. In 3nm technology, the super fin will provide a fivefold increase in capacitance within the same footprint, resulting in a voltage reduction and substantially enhanced product performance.

In the future, an extremely powerful computer architecture confinement may be achieved by merging very large-scale integration (VLSI) technology and AI. The researchers are using AI approaches in VLSI design automation to solve difficulties at various design stages. Knowledge-based and expert systems are AI approaches that attempt to describe the issue first and then select the best solution from a set of alternatives. AI languages are well suited to tackling issues of this complexity [20] The Internet of Things (IoT) is a sophisticated automation and analytics system that uses networking, sensors, big data, and machine learning to create full systems for a product or service. When applied to any industry or system, these systems provide improved transparency, control, and performance. Users may accomplish greater automation, research, and integration inside a system using IoT solutions. The availability of low-cost VLSI devices is essential for the IoT's success. As the IoT grows, we can identify various potential problems in the design of VLSI IoT devices [21]. In the future, when IoT is clubbed with 3nm technology, the design of IoT devices will provide low-cost and low-power consumption devices. In system-in-package designs, integrating low-power operation, sensing, computing, and communication frequently necessitate merging innovative technologies with mature production nodes. Low-cost IoT systems encourage the use of tiny chips that may be deployed as part of bigger networks [21]. The cost of silicon is simply one element in determining the IoT's ultimate acceptability [22].

As we scale down these chips, they also find application in wireless communications like 5G and 6G. We will have devices with better data rates, capacity, coverage, and low latency, as well as devices that are less expensive, energy-efficient, and more durable than previous technologies. Technological advances in VLSI are projected to expand the scope, efficiency, and effectiveness of data-handling processes while also lowering costs. Device control, data collection, information processing and management, and information display tasks will be partitioned into units that may be implemented with separate processors and memory to satisfy particular functional needs in future office systems and subsystems. Workstations, printers, storage units, and communication facilities will all have dedicated CPUs [23].

REFERENCES

1. Arora, Sneha, Umesh Dutta, and Vipin Kumar Sharma, "A Noise Tolerant and Low Power Dynamic Logic Circuit Using Finfet Technology", International Journal of Engineering Research and Applications (IJERA), Vol. 5, No. 12, December 2015, ISSN: 2248-9622. Available at: www.ijera.com
2. Maszara, W.P., and M. Lin, "FinFETs—Technology and Circuit Design Challenges," in 2013 Proceedings of the ESSCIRC (ESSCIRC), 2013, PP. 3–8. https://doi.org/10.1109/ESSCIRC.2013.6649058.
3. Barraud, S., et al., "Performance and Design Considerations for Gate-all-around Stacked-Nano Wires FETs", 2017 IEEE International Electron Devices Meeting (IEDM), 2017, PP. 29.2.1–29.2.4. https://doi.org/10.1109/IEDM.2017.8268473.
4. Alam, S., A. Raman, B. Raj, N. Kumar, and S. Singh, "Design and Analysis of Gate Overlapped/Underlapped NWFET Based Lable Free Biosensor", Silicon, PP. 1–8, 2021.
5. Karbalaei, M., D. Dideban, and H. Heidari, "A Sectorial Scheme of Gate-all-around Field Effect Transistor with Improved Electrical Characteristics", Ain Shams Engineering Journal, Vol. 12, No. 1, PP. 755–760, 2021.
6. Alam, S., A. Raman, B. Raj, and N. Kumar, "Design and Analysis of Gate Underlapped/Overlapped Surround Gate Nanowire TFET for Analog Performance", in 2019 4th International Conference on Recent Trends on Electronics, Information, Communication & Technology (RTEICT). IEEE, May 2019, PP. 454–458.
7. Sung-Young, Lee, et al., "A Novel Multibridge-channel MOSFET (MBCFET): Fabrication Technologies and Characteristics", IEEE Transactions on Nanotechnology, Vol. 2, No. 4, PP. 253–257, December 2003. https://doi.org/10.1109/TNANO.2003.820777.
8. Bae, G., et al., "3nm GAA Technology Featuring Multi-Bridge-Channel FET for Low Power and High Performance Applications", in 2018 IEEE International Electron Devices Meeting (IEDM), 2018, PP. 28.7.1–28.7.4. https://doi.org/10.1109/IEDM.2018.8614629.
9. Singh, S., and B. Raj, "Analytical Modeling and Simulation Analysis of T-shaped III-V Heterojunction Vertical T-FET", Superlattices and Microstructures, Elsevier, Vol. 147, PP. 106717, November 2020.
10. Chawla, T., M. Khosla, and B. Raj, "Optimization of Double-gate Dual Material GeOI-Vertical TFET for VLSI Circuit Design", IEEE VLSI Circuits and Systems Letter, Vol. 6, No. 2, PP. 13–25, August 2020.
11. Kaur, M., N. Gupta, S. Kumar, B. Raj, and Arun Kumar Singh, "RF Performance Analysis of Intercalated Graphene Nanoribbon Based Global Level Interconnects", Journal of Computational Electronics, Springer, Vol. 19, PP. 1002–1013, June 2020.
12. Wadhwa, G., and B. Raj, "An Analytical Modeling of Charge Plasma Based Tunnel Field Effect Transistor with Impacts of Gate Underlap Region", Superlattices and Microstructures, Elsevier, Vol. 142, PP. 106512, June 2020.
13. Singh, S., and B. Raj, "Modeling and Simulation Analysis of SiGe Hetrojunction Double Gate Vertical T-shaped Tunnel FET", Superlattices and Microstructures, Elsevier, Vol. 142, P. 106496, June 2020.

14. Singh, S., and B. Raj, "A 2-D Analytical Surface Potential and Drain Current Modeling of Double-Gate Vertical T-shaped Tunnel FET", Journal of Computational Electronics, Springer, Vol. 19, PP. 1154–1163, April 2020.
15. Singh, S., S. Bala, B. Raj, and B. Raj, "Improved Sensitivity of Dielectric Modulated Junctionless Transistor for Nanoscale Biosensor Design", Sensor Letter, ASP, Vol. 18, PP. 328–333, April 2020.
16. Kumar, V., S. Kumar, and B. Raj, "Design and Performance Analysis of ASIC for IoT Applications", Sensor Letter, ASP, Vol. 18, PP. 31–38, January 2020.
17. Wadhwa, G., and B. Raj, "Design and Performance Analysis of Junctionless TFET Biosensor for High Sensitivity", IEEE Nanotechnology, Vol. 18, PP. 567–574, 2019.
18. Wadhera, T., D. Kakkar, G. Wadhwa, and B. Raj, "Recent Advances and Progress in Development of the Field Effect Transistor Biosensor: A Review", Journal of Electronic Materials, Springer, Vol. 48, No. 12, PP. 7635–7646, December 2019.
19. Singh, S., and B. Raj, "Design and Analysis of Hetrojunction Vertical T-shaped Tunnel Field Effect Transistor", Journal of Electronics Material, Springer, Vol. 48, No. 10, PP. 6253–6260, October 2019.
20. Goyal, C., J.S. Ubhi, and B. Raj, "A Low Leakage CNTFET Based Inexact Full Adder for Low Power Image Processing Applications", International Journal of Circuit Theory and Applications, Wiley, Vol. 47, No. 9, PP. 1446–1458, September 2019.
21. Sharma, S.K., B. Raj, and M. Khosla, "Enhanced Photosensivity of Highly Spectrum Selective Cylindrical Gate In1-xGaxAs Nanowire MOSFET Photodetector", Modern Physics Letter-B, Vol. 33, No. 12, P. 1950144, 2019.
22. Singh, J., and B. Raj, "Design and Investigation of 7T2M NVSARM with Enhanced Stability and Temperature Impact on Store/Restore Energy", IEEE Transactions on Very Large Scale Integration Systems, Vol. 27, No. 6, PP. 1322–1328, June 2019.
23. Bhardwaj, A.K., S. Gupta, B. Raj, and Amandeep Singh, "Impact of Double Gate Geometry on the Performance of Carbon Nanotube Field Effect Transistor Structures for Low Power Digital Design", Computational and Theoretical Nanoscience, ASP, Vol. 16, PP. 1813–1820, 2019.
24. Goyal, C., J. Subhi, and B. Raj, "Low Leakage Zero Ground Noise Nanoscale Full Adder Using Source Biasing Technique", Journal of Nanoelectronics and Optoelectronics, American Scientific Publishers, Vol. 14, PP. 360–370, March 2019.
25. Singh, A., M. Khosla, and B. Raj, "Design and Analysis of Dynamically Configurable Electrostatic Doped Carbon Nanotube Tunnel FET", Microelectronics Journal, Elesvier, Vol. 85, PP. 17–24, March 2019.
26. Goyal, C., J.S. Ubhi, and B. Raj, "A Reliable Leakage Reduction Technique for Approximate Full Adder with Reduced Ground Bounce Noise", Journal of Mathematical Problems in Engineering, Hindawi, Vol. 2018, PP. 16, Article ID 3501041, 15 October 2018.
27. Wadhwa, G., and B. Raj, "Label Free Detection of Biomolecules Using Charge-Plasma-Based Gate Underlap Dielectric Modulated Junctionless TFET", Journal of Electronic Materials (JEMS), Springer, Vol. 47, No. 8, PP. 4683–4693, August 2018.

28. Wadhwa, G., and B. Raj, "Parametric Variation Analysis of Charge-Plasma-based Dielectric Modulated JLTFET for Biosensor Application", IEEE Sensor Journal, Vol. 18, No. 15, 1 August 2018.
29. Yadav, D., S.S. Chouhan, S.K. Vishvakarma, and B. Raj, "Application Specific Microcontroller Design for IoT based WSN", Sensor Letter, ASP, Vol. 16, PP. 374–385, May 2018.
30. Singh, G., R.K. Sarin, and B. Raj, "Fault-Tolerant Design and Analysis of Quantum-Dot Cellular Automata Based Circuits", IEEE/IET Circuits, Devices & Systems, Vol. 12, PP. 638–664, 2018.
31. Singh, J., and B. Raj, "Modeling of Mean Barrier Height Levying Various Image Forces of Metal Insulator Metal Structure to Enhance the Performance of Conductive Filament Based Memristor Model", IEEE Nanotechnology, Vol. 17, No. 2, PP. 268–267, March 2018 (SCI).
32. Jain, A., S. Sharma, and B. Raj, "Analysis of Triple Metal Surrounding Gate (TM-SG) III-V Nanowire MOSFET for Photosensing Application", Opto-electronics Journal, Elsevier, Vol. 26, No. 2, PP. 141–148, May 2018.
33. Jain, N., and B. Raj, "Parasitic Capacitance and Resistance Model Development and Optimization of Raised Source/Drain SOI FinFET Structure for Analog Circuit Applications", Journal of Nanoelectronics and Optoelectronins, ASP, USA, Vol. 13, PP. 531–539, April 2018.
34. Lee, Sung-Young, et al., "A Novel Sub-50 nm Multi-bridge-channel MOSFET (MBCFET) with Extremely High Performance", Digest of Technical Papers. 2004 Symposium on VLSI Technology, 2004, PP. 200–201. https://doi.org/10.1109/VLSIT.2004.1345478.
35. Oh, Chang-Woo, et al., "Highly Manufacturable Sub-50 nm High Performance CMOSFET Using Real Damascene Gate Process", 2003 Symposium on VLSI Technology. Digest of Technical Papers (IEEE Cat. No. 03CH37407), 2003, PP. 147–148. https://doi.org/10.1109/VLSIT.2003.1221128.
36. Kim, S.M., C.W. Oh, J.D. Choe, C.S. Lee, and D.G. Park, "A Study on Selective Si0. 8Ge0. 2 Etch Using Polysilicon Etchant Diluted by H2O for Three-dimensional Si Structure Application", in *Proceedings-Electrochemical Society*, 2003, PP. 81–86.
37. Cheng, Y., and C. Hu, MOSFET Modeling & BSIM3 User's Guide. Springer Science & Business Media, 1999.
38. Chang, L., K.J. Yang, Y.C. Yeo, I. Polishchuk, T.J. King, and C. Hu, "Direct-Tunneling Gate Leakage Current in Double-gate and Ultrathin Body MOSFETs", IEEE Transactions on Electron Devices, Vol. 49, No. 12, PP. 2288–2295, 2002.
39. Joung, S., and S. Kim, "Leakage Performance Improvement in Multi-Bridge-Channel Field Effect Transistor (MBCFET) by Adding Core Insulator Layer", in 2019 International Conference on Simulation of Semiconductor Processes and Devices (SISPAD). IEEE, September 2019, PP. 1–4.
40. Sepranos, D., and M. Wolf, "Challenges and Opportunities in VLSI IoT Devices and Systems", IEEE Design & Test, Vol. 36, No. 4, PP. 24–30, August 2019. https://doi.org/10.1109/MDAT.2019.2917178.
41. Wolf, M., and S. Mukhopadhyay, "VLSI for the Internet of Things", Computer, Vol. 50, No. 6, PP. 16–18, 2017. https://doi.org/10.1109/MC.2017.158.
42. Einspruch, Norman G., "Chapter 8—The Effect of VLSI: Prospects for the Office of the Future", VLSI Electronics Microstructure Science, Elsevier, Vol. 4, PP. 243–282.

Chapter 5

High-Speed Nanoscale Interconnects

Somesh Kumar and Manoj Kumar Majumder

CONTENTS

In this chapter, we discuss the important role played by high-speed on-chip, that is, nanoscale interconnects in the design of integrated circuits and systems. Losses and different types of interconnects, that is, graphene-based interconnects and hybrid interconnects, are also described. The future predictions are also presented at the end of this chapter.

DOI: 10.1201/9781003311379-5

5.1 INTRODUCTION OF INTERCONNECTS

A chip, or an integrated circuit (IC), is a set of electronic components and circuits placed on a silicon substrate, and these electronic circuits and components are electrically connected to each other with the help of planar or vertical conductors called interconnects. Interconnects are primarily fabricated using metals with high-conductive properties and form the backbone for signal communication carrying data and clock signals on a chip. The design of interconnects on an IC is vital to its functionality, performance, reliability, efficiency, and overall fabrication yield. The material of interconnects is decided considering many factors, such as chemical and mechanical compatibility, electrical conductivity, and fabrication challenges [1]. The performance of ICs is increasingly dependent on the overall signal, power, and thermal integrity of these high-speed interconnects.

Robert Noyce at Fairchild Semiconductors invented the monolithic ICs with aluminum (Al) over Si as the interconnect material in 1959. From 1959 to the early 1990s, Al has been widely used as an interconnect material due to its high conductivity, ease of deposition, and excellent adhesion properties [2]. Tungsten has also been popularly used for interconnects; however, it has a higher resistivity than that of Al. The electromigration issue in Al-based interconnects reduces the lifetime. It degrades the reliability of ICs [3]. Due to this reason, in the early 2000s, Copper (Cu) interconnects replaced the Al-based interconnects due to their higher conductivity and lower electromigration when compared to Al [4]. Typically, in the front-end-of-line, discrete components, that is, transistors, capacitors, diodes, and others, are fabricated on the wafer, and these components are interconnected to transmit information (data and clock) to the backend-of-line. Many interconnects are needed in complex ICs, and it is impossible to create all these connections in a single layer on top of the wafer [5]. Therefore, chip manufacturers have built multiple levels of interconnects for proper routing and signaling [6]. Simple ICs may have just one or two metal layers, while complex ICs can have ten or more layers of wiring. Interconnect metal layers are interconnected vertically by etching holes that are called vias.

Typically, interconnects are broadly classified into on-chip interconnects and chip-to-chip interconnects [7, 8]. On-chip Cu interconnects are used to distribute signals/clocks and provide power/ground to the various components/subsystems on the IC. Based on the level of interconnects, there are three types of on-chip interconnects—local, intermediate, or semi-global—and global interconnects, as shown in Figure 5.1. There is a constant need for small and thin interconnects near the transistors because of their smaller sizes and dense packaging. Normally, these first-level lines are called local interconnects, which consist of deeply scaled lines. In the metal–oxide semiconductor (MOS) technology, source/drains, and gates are physically connected using local interconnects, and bases, emitters, and collectors in bipolar junction transistors within a functional block.

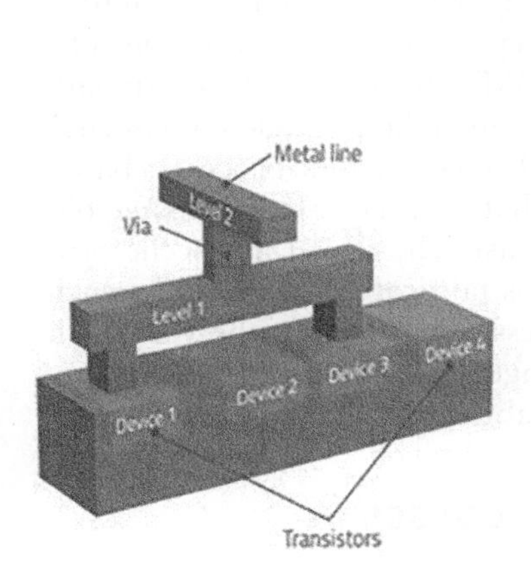

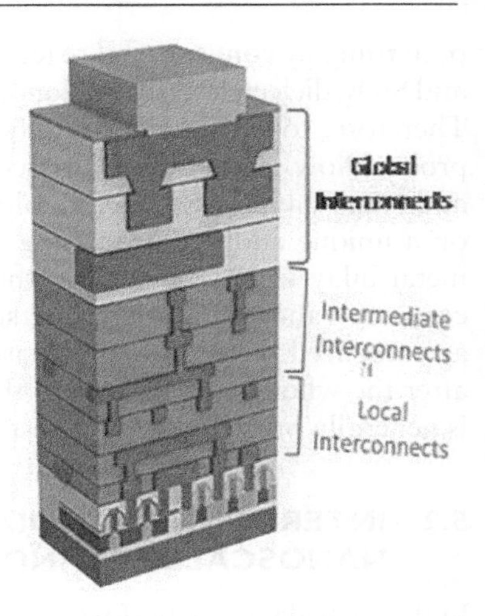

Figure 5.1 Multiple levels of interconnects [5, 7].

Intermediate or semi-global Cu interconnects are thicker and longer than the local interconnects, resulting in lower resistance. Semi-global interconnects are primarily used for clock and signal distribution networks within a functional block with typical lengths of the order of a few millimeters. Global interconnects are thicker and taller (>4mm) than intermediate interconnects and provide inter-block clock or signal distribution. Global interconnects are mostly used to deliver power/ground to all functional blocks and occupy some of the top layers, as shown in Figure 5.1 [9]. Local interconnects usually occupy the first few metal layers and provide higher resistivity than semi-global/intermediate or global wires due to their shorter lengths.

In the 1970s and 1980s, devices with one or two layers of sputtered Al alloy had a size of approximately 1μm during their mass production. For local interconnections and gate electrodes, poly-Si was used. To reduce the gate resistance, a sandwich of silicide and poly-Si was used. Local planarization, Al alloys, silicide contacts, and poly-SI gates were used in the ICs. However, by the 1990s, silicide chemical vapor deposition (CVD), global planarization with tungsten plugs, and low-k dielectrics were prevalent in ICs. In 1995, Motorola Inc. fabricated a microprocessor with five Al layers. However, by the late 1990s, Al was getting replaced by Cu because of higher conductivity, longer lifetime, fewer hillocks, and lower resistance-capacitance (RC) delay [10]. Today, technology advancements primarily lead toward ten times smaller interconnections that allow up to 15 levels of metallization to be laid on the chip.

During the transition to Cu interconnects, researchers faced various challenges with respect to their fabrication. These included difficulty in the patterning by conventional reactive ion etching (RIE), diffusion of Cu into Si and SiO_2, dielectric and junction leakage, and poor corrosion/oxidation [11]. Therefore, to tackle all these challenges, IBM introduced a revolutionary process flow named the Damascene process, including "single-damascene" and "dual-damascene" methodologies [11]. The Damascene process is based on a unique additive processing principle that originates from the age-old metal inlay technique used in the Middle East. The dual-Damascene process is advantageous due to the simultaneous fabrication of via and trench as compared to the single damascene process that develops the via/trench after the whole process. Due to this advantage, dual damascene technology is generally preferred for the cost-effective fabrication of Cu interconnects.

5.2 INTERCONNECT MODELS AND LOSSES IN NANOSCALE TECHNOLOGY

In the early days, the performance of interconnects was not of much interest due to the lower operating speeds of microwave integrated circuits. However, the demand for higher speeds, increased integration density, and the miniaturization of devices played a key role in focussing our attention to interconnect performance [12, 13]. In the early phase of semiconductor technology, the interconnect capacitance alone determined the overall delay of an IC. A few years later, miniaturization of dimensions, increased complexity and use of longer interconnects forced the designers to consider resistive components as well [14]. However, the simple RC model soon became overly optimistic at higher operating speeds [15]. For modern ICs at lower technology nodes, the inductive effect also plays a major role in determining the performance of interconnects. The electrical parameters, namely, resistance (R), capacitance (C), and inductance (L), can be used to describe any interconnect model. Furthermore, the type of material and the geometry of interconnects are primarily responsible for their quantitative behavior. The primary reason behind modeling interconnects is to verify the signal integrity and provide detailed insight of the physical properties of the wire [16, 17]. Therefore, a suitable electrical model is absolutely required for analyzing the interconnects. The equivalent electrical network of interconnects is categorized as lumped and distributed capacitance/resistance/RC/RLC transmission-line based interconnect models. The different local, intermediate, and global interconnect structures, along with lumped and distributed RLC models, are demonstrated in Figure 5.2.

The transistor scaling, integration, and cost are reduced dramatically in next-generation ICs by following Moore's predictions [18]. Moore's law for ICs has brought the semiconductor industry to a net worth of more than a trillion dollars. However, Moore's law, which includes both reductions in cost and increases in transistor density in nearly two years, is predicted to come to an end as the current technology nodes approach the atomistic

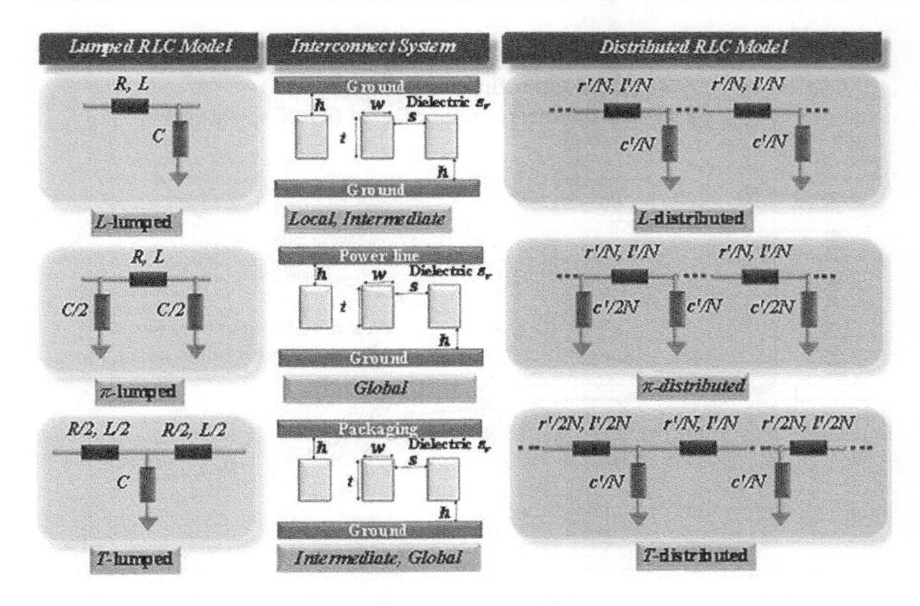

Figure 5.2 Interconnect structures and different *RLC* models.

dimensions [19]. In the past 50 years, the manufacturing industry has also been developed and equipped with advanced technology to develop high-speed devices. The very large-scale integration (VLSI) industry has been accomplished using these advanced units to scale the feature size from 10 µm down to 10 nm. In order to meet future requirements, the devices and interconnects are aggressively scaled just after changes in advanced technology by integrating millions of transistors on ICs or multiple systems on a chip. Thus, their performance has increased due to systematic transistor scaling. However, interconnects continue to face a significant roadblock to further scaling as compared to devices [20]. The expansion of the equivalent *RC* delay is the major cause of an increase in the interconnect losses and signal integrity; however, the transistor delay decreases due to scaling, as depicted in Figure 5.3. It is, therefore, apparent that the interconnect delay cannot be neglected for next-generation technology due to the higher dominating factor of interconnect as compared to the delay of a transistor [21, 22]. Therefore, interconnect performance is imperative to investigate more rigorously for low loss and high-speed devices and interconnects.

The dimensions and materials are some of the significant factors of the interconnect that determine the signal integrity. With advances in digital circuit technologies over the last few decades, on-chip interconnects are increasingly seen as the limiting factor on the performance and power budget of a chip. Thus, any improvement in on-chip interconnects can lead to

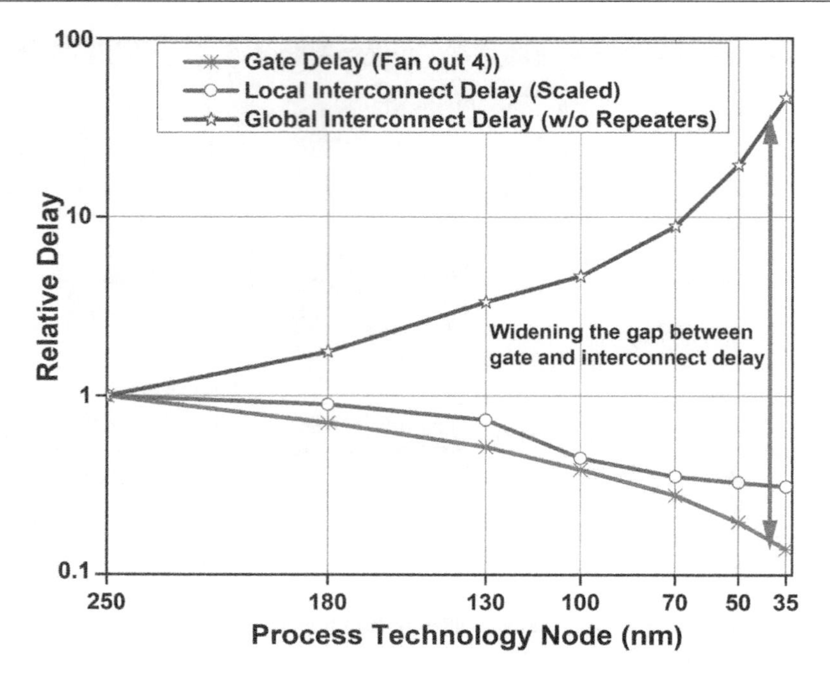

Figure 5.3 Impact of scaling on the gate and interconnect delays [21].

overall progress on ICs' performance. However, the performance of the chip cannot be claimed to provide enhancements in the performance of the entire system. For example, the performance is not only dependent on the computation performed based on the faster microprocessor but also dependent on the different level of cache or main memory involved in the process to fetch the data or information. Therefore, if the synchronization cannot be maintained between the on-chip bandwidth from the main memory to the microprocessor, the overall computation process will not provide improved performance as expected [23]. Furthermore, the main concern of the designer has shifted from the optimization of logic to interconnects due to the dominating factor of interconnect delay at high operating frequency and lower physical dimensions. In this regard, several performance metrics such as lower power dissipation, noise, physical area, energy-delay product (EDP), and signal integrity, among others, need to be considered before proceeding for the further design of interconnects. Therefore, the performance of interconnects primarily depends on these metrics, which mainly rely on the interconnect parasitics, such as L, R, and C [24]. Furthermore, the interconnect parasitics are more sensitive and vary significantly with the miniaturization of technology, causing degradation in performance for on-chip Cu interconnects [7, 8]. Thus, one needs to clearly represent the interconnect line in terms of an

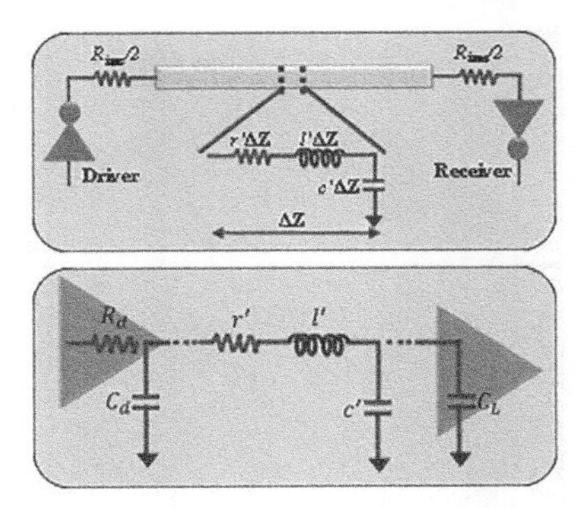

Figure 5.4 (a) Equivalent distributed circuit of interconnect and (b) a DIL arrangement of interconnects.

equivalent circuit, as shown in Figure 5.4. Figure 5.4a represents the equivalent circuit model for a single conductor line. The wire is represented by a distributed *RLC* transmission line, where *l*, *r* and *c* are the per unit length (*p.u.l.*) L, R, and C, respectively. Figure 5.4b shows the driver-interconnect-load (DIL) system used for the analysis of interconnects. The C_d and R_d are the source capacitance and resistance, respectively, and C_L denotes the capacitive load.

Based on the International Technology Roadmap for Semiconductors (ITRS), to achieve the high data rates and low energy-per-bit necessities, the total losses, including conductor and dielectric losses at current and future technology nodes, need to be reduced. The conductor loss in Cu interconnects is primarily classified into two components: DC loss and frequency-dependent loss. DC loss mainly depends on the cross-sectional area, length, and resistivity of the signal conductor through which the current is flowing. In advanced technology and by miniaturizing the feature size, Cu exhibits higher resistivity due to surface roughness, grain boundary, and surface scattering [7, 8]. The following factors have a substantial impact on the overall Cu losses.

5.2.1 Size Effect in On-Chip Cu Interconnects

For sub-100*nm* technology nodes in on-chip interconnects, when the feature size of the interconnects becomes comparable to the bulk mean free path (MFP) of copper (λ_{bulk}), that is, 39 *nm*, the resistivity increases drastically due to electron, grain, and surface boundary scattering effects [7, 8, 25]. This phenomenon is called the size effect and occurs due to the surface and grain boundaries formed during the deposition of the Cu. The surface and grain-boundary scattering phenomena are shown in Figure 5.5.

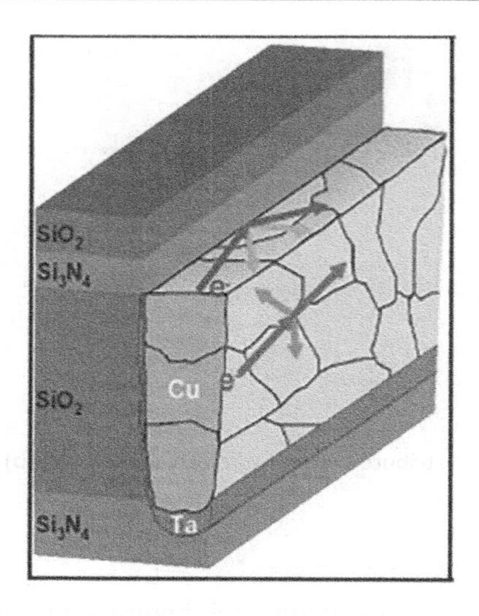

Figure 5.5 Grain boundary and surface scattering of electrons [25].

The effect of surface scattering was the first given by Fuchs and later Sondheimer extended this for narrow metal lines. Collectively, the Fuch–Sondheimer (FS) model demonstrates their surface scattering phenomenon. The FS model is derived from Boltzmann transport equations and is given by [26, 27]:

$$\rho_s = \frac{\rho_{bulk}}{1 - \frac{3(1-p)}{4k}\int_0^1 dx(x - x^3)\frac{\left(1 - e^{-k/x}\right)}{\left(1 - pe^{-k/x}\right)}}, \tag{5.1}$$

where $k = t/\lambda_0$, is the ratio of the thickness to the electron bulk MFP, ρ_{bulk} is the bulk resistivity of the conductor line, p is the specularity parameter used to describe the fraction of the electrons that are elastically scattered at the surface and their values lie between 0 to 1, t is interconnect thickness, and the constant x is dependent on the scattering angle (θ) and is given as $x = cos\theta$. Grain boundaries in a polycrystalline interconnect act like partially reflecting planes, as illustrated in Figure 5.5. Grain sizes are usually scaled linearly with the wire dimensions, and when the grain size is comparable to the electron MFP, the electrons grain-boundary scattering increases, which further increases the conductor resistivity. The effect of grain-boundary scattering is given by the Mayadas–Shatzkes (MS) model and is given as [28].

$$\rho_g = \frac{\rho_{bulk}}{3}\left[\frac{1}{3} - \left(\frac{\lambda_0}{2d}\frac{\chi}{(1-\chi)}\right) + \left(\frac{\lambda_0}{d}\frac{\chi}{(1-\chi)}\right)^2 - \left(\frac{\lambda_0}{d}\frac{\chi}{(1-\chi)}\right)^3 \times \ln\left(1 + \frac{d(1-\chi)}{\lambda_0 \chi}\right)\right]^{-1}, \tag{5.2}$$

where χ and d is termed reflection coefficient and the average grain size, respectively.

5.2.2 Effect of Temperature on Resistivity

A linear relationship between the resistivity of copper and temperature [29] can be expressed as

$$\rho_t = \left[1 + \alpha(T - T_0)\right]\rho_{bulk}. \tag{5.3}$$

Here, T_0, α, and ρ_{bulk} are the room temperature, the temperature coefficient of resistance, and the bulk resistivity at room temperature, respectively. It is observed that the MFP of the Cu decreases with an increase in the temperature, leading to an increase in resistivity. The individual parameters λ_0 and ρ_{bulk} are temperature-dependent, while the product of these is a temperature-independent constant.

5.2.3 Surface Roughness

While investigating electrical losses, it is generally anticipated that the conductor material possesses smooth outer surfaces and homogeneous nature. This is not the actual fact as Cu is usually roughed to increase adhesion to the dielectric laminate. Thus, these rough conductor profiles yield conductor losses at high frequencies [7, 8, 29].

Conductor surface roughness effects refer to the influence of horizontal and vertical grooves on the surface of the current flow in a conductor and the subsequent change in the electromagnetic properties of the conductor. The surface roughness in Cu lines produces nonuniform structures with dissimilar spatial distances and heights. It is observed that the scattering phenomenon occurs more in the rough surfaces of Cu lines in the nanometer range that yields to size effects, resulting in a lower effective MFP, a higher effective resistivity, and a global increase in resistance [7]. In addition, the current passing through rough surfaces yields the creation of magnetic loops over the rough surfaces that leads to a rise in the inductance. Furthermore, surface roughness causes an additional conductor surface area so as to enhance the parasitic capacitances [7, 8]. Thus, there is an overall penalty

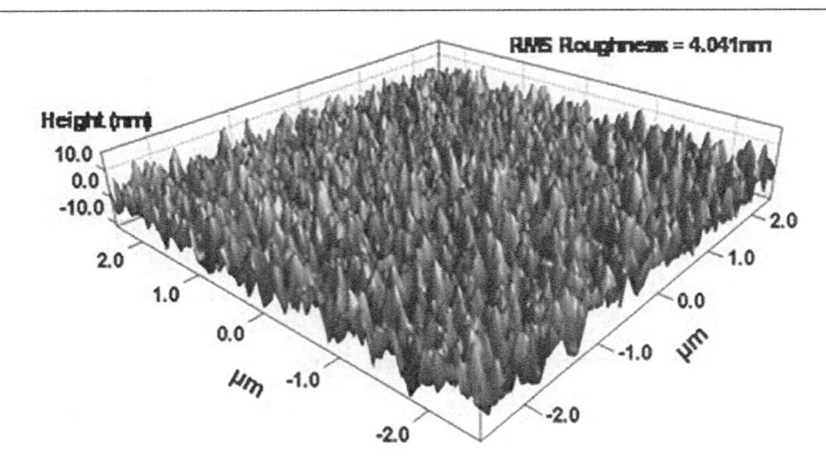

Figure 5.6 A 16*nm* thin sputtered Cu film grown on Si (100) substrate (scan area of 5μm×5μm) [7].

on the line parasitics due to the surface roughness. This can further enhance the penalty on performance metrics such as energy, delay, signal integrity, electronic data processing (EDP), and bandwidth density (BWD). Figure 5.6 shows the typical atomic force microscope (AFM) scan of 16*nm*-thick Cu sheet with the scan sizes 5μm × 5μm. Using this AFM analysis, the roughness parameters can be extracted.

For a smooth and rough line, the FS-MS effective resistivity model is shown in Figure 5.7 as a function of technology. For a line width ranging from 45*nm* to 10*nm*, the effective resistivity of local/intermediate lines is presented for the different extent of roughness. It is evident that the total resistivity of the interconnect is a function of the surface roughness.

5.2.4 Current Density

Cu at sub-22-*nm* technology nodes exhibit a current density of 10^6 A/cm^2 that has forced designers to look for alternative materials that can meet the foreseeable challenges in ultra-scaled interconnects [31]. Replacing or improving the current Cu interconnects is becoming imminent as line widths continue to shrink. In this context, graphene has emerged as the prime material of choice for interconnects due to its low resistivity and susceptibility to high-current density.

As shown in Figure 5.8, while the maximum current density (red dotted line) is shrinking, the more marked decrease in cross-sectional areas (green line) leads to increasing current densities [32]. Second, smaller interconnects require higher current densities to perform their intended functionality (blue dotted line), while at the same time tolerable current-density limits are shrinking (green dotted line).

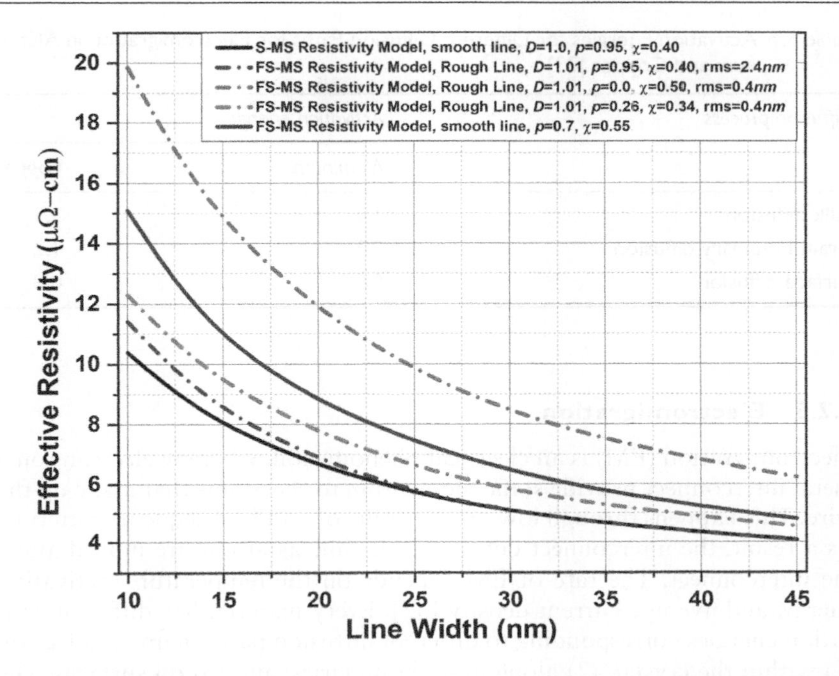

Figure 5.7 FS-MS resistivity model for the smooth and rough line considering 300K temperature and an equal average grain size of w [30].

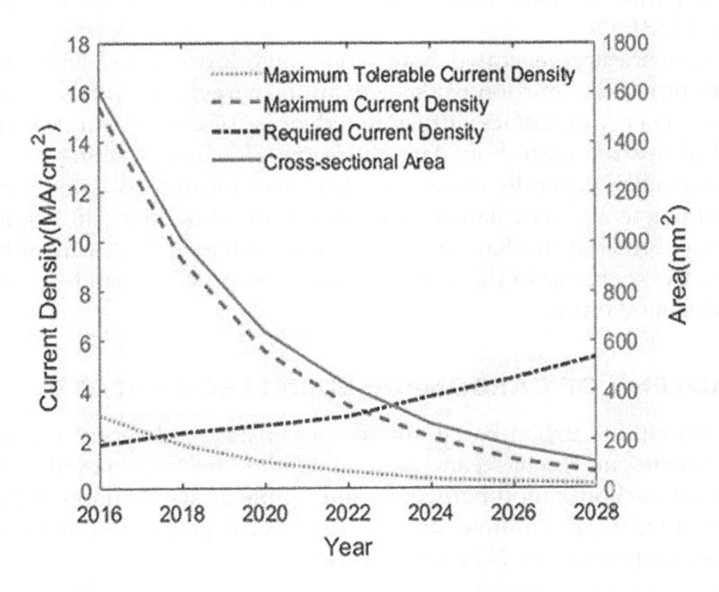

Figure 5.8 Current density and cross-sectional area projection for Cu based on the ITRS.

Table 5.1 Activation Energies for Different Diffusion Paths for Electromigration in Aluminum and Copper [33]

Diffusion process	*Activation energy*	
	Aluminum	*Copper*
Bulk diffusion	1.2	2.3
Grain-boundary diffusion	0.7	1.2
Surface diffusion	0.8	0.8

5.2.5 Electromigration

Electromigration (EM) is an electrical phenomenon, whereby electrons on a metal interconnect provide some momentum to the atoms that make up the wire. This happens through low-energy collisions and subsequent scattering. As a result, the interconnect deforms over time as atoms are moved along the interconnect. The rate of EM depends on the temperature, activation energy, and average current density [32]. Every material has different activation energies corresponding to different diffusion paths, namely, diffusion (1) within the crystal, (2) along grain boundaries, and (3) on surfaces. The diffusion mechanism depends on activation energy. Surface scattering and grain boundary are dominant diffusion mechanisms in Cu; to stop the diffusion of Cu into the dielectric material, a barrier layer is employed. Different diffusion paths are characterized by different activation energies E_a given in Table 5.1 [33].

The temperature generated both from Joule heating and ambient temperature provides a portion of the activation energy as thermal energy. The activation energy of surface diffusion in the case of copper can be increased as well above the grain-boundary diffusion. However, curbing one diffusion mechanism generally causes another mechanism to become predominant, leading to alternate damage scenarios. Grain-boundary diffusion is the dominant diffusion mechanism in nanoscale interconnects with diffusion barriers. Every change in the dominant diffusion process, therefore, changes the failure modalities.

5.3 ADVENT OF CARBON-BASED INTERCONNECTS

The research in carbon-based interconnects has undergone remarkable advancements in recent years. Several studies have proposed different models for carbon nanotube (CNT) and graphene nanoribbon (GNR) by incorporating unique atomic structures, physical properties, and electron transport properties of CNTs and GNRs.

5.3.1 Evolution of Technology

Recently, the semiconductor industry reached its maximum revenue of 402 billion dollars in 2020, as mentioned by the World Semiconductor Trade Statistics (WSTS) [34], and the technical progress is exemplified by leading electronic products. While Cu exhibits ten times more immunity to fight against the electromigration effect, better reliability, and high-current density than conventional materials. However, the rapid scaling of interconnects in the semiconductor industry limits the functionality of Cu for interconnect applications [6, 35]. At the nanometre technology node, Cu interconnects suffer from several reliability and electrical issues that have limited the further scaling of interconnects. Therefore, we are forced to find alternative solutions in the form of emerging carbon-based interconnect materials, such as CNTs, hybrid, and GNRs.

5.3.2 CNTs' Unique Structures and Types

Norio Taniguchi first proposed the term *nanotechnology* in 1974 as a precise way to highlight the physical dimensions on ultra-scaled circuits, devices, and systems [36]. In 1981, the scanning tunneling microscope (STM) invention opened a new gateway to understanding the material properties in a detailed manner. The invention of CNTs by Sumio Iijima in 1991 accelerated the new nanotechnology era, with CNTs being considered a suitable material for interconnect applications [36, 37]. In 2003, Konstantine Novoselov and Andre Geim changed the electronic industry's whole perspective after revealing their research by a simple scotch tap method [38]. For their groundbreaking experiments on a one-atom-thick layer of graphite, they were awarded the Nobel Prize in Physics in 2010 [38]. The nature of CNTs is primarily dependent on the way one-atom-thick graphene sheet is rolled up to produce nanotubes [38]. However, the growth of CNTs in a realistic scenario does not incorporate the rolled-up phenomenon of the graphene sheet. Instead, it can be produced by using several CVD processes that are primarily responsible for the direct growth of CNTs vertically over the substrate. The physical properties of CNTs can be distinguished by varying the diameter, *dia*, angle θ (in the range of 0–30 degrees), and pair of chiral indices (n, m). The different scenario of CNTs that exhibits armchair, zigzag, and chiral structures can be obtained based on the analysis demonstrated in Figure 5.9. The pair of chiral indices (n, m) and diameter of CNTs are primarily responsible for the nature of the structure, which can be semiconducting or metallic. The chiral indices (n, m) for armchair structure exhibits metallic behavior only when n and m values are equal, whereas for zigzag structure, one of the indices values must be zero (either $n = 0$ or $m = 0$) and it must satisfy $n - m = 3i$ (where i is a positive integer) condition to behave as metal or semiconductor. Depending on the number of concentric rolled

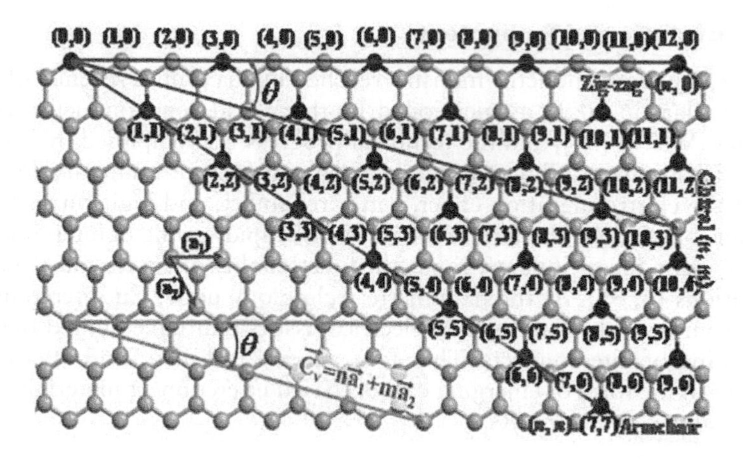

Figure 5.9 A honeycomb lattice structure of graphene along with electronic patterns and unit cells.

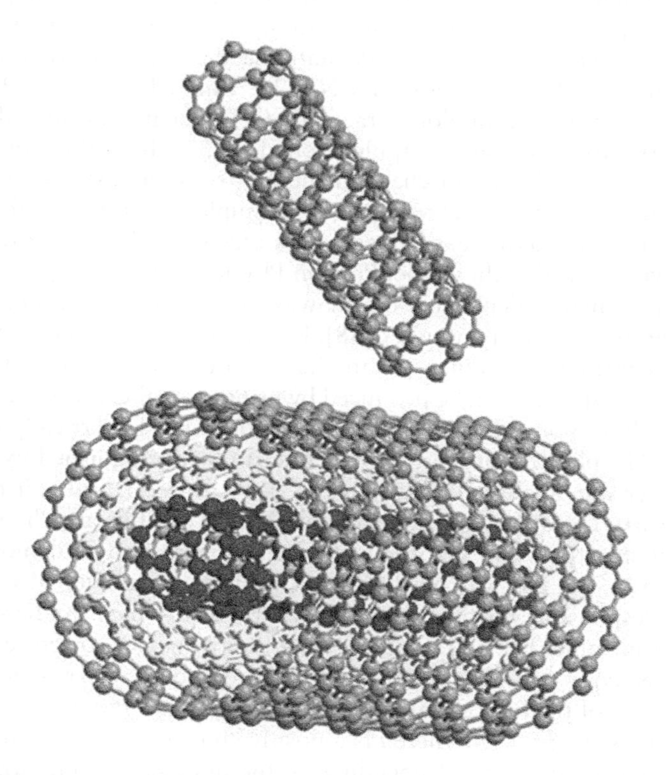

Figure 5.10 Structures of (a) SWCNT and (b) MWCNT.

cylindrical tubes, CNTs can be categorized as single- and multiwalled CNT (SWCNT and MWCNT, respectively) as demonstrated in Figures 5.10a and 5.10b, respectively. If the nanotubes contain a single cylindrical wall with diameters ranging from 0.7–4.33*nm*, it is termed as SWCNT; however, CNTs with more than one cylindrical wall with a separation of interlayer Vander wall distance of 0.34*nm* can be treated as an MWCNT.

5.3.3 GNR Structures and Types

A strip of graphene having a width less than 50*nm* can be defined as a GNR that can be armchair and zigzag depending on the arrangement of carbon atoms on the edges as depicted in Figures 5.11a and 5.11b, respectively. The theoretical model of GNR was efficiently demonstrated by Mitsutaka Fujita research group in 1996. The analysis was carried out to demonstrate the impact of the edge and scaling effect on GNRs. Later, scientists Konstantin Novoselov and Andre Geim in 2004 succeeded in obtaining a one-atom-thick monolayer of carbon for the first time; later, the similar research group

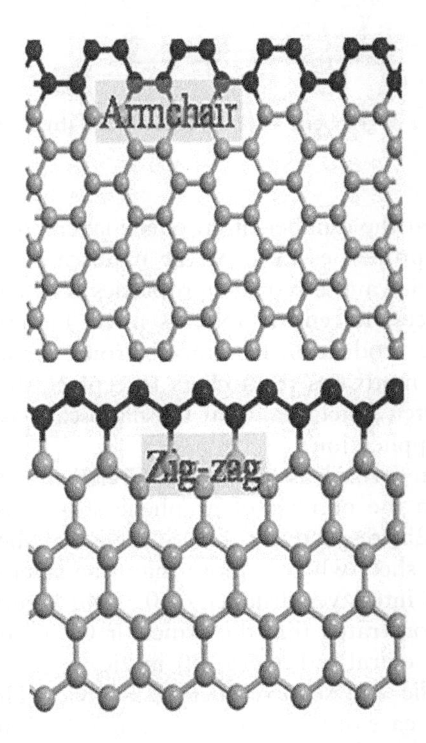

Figure 5.11 Two-dimensional view of (a) an armchair and (b) a zigzag GNR.

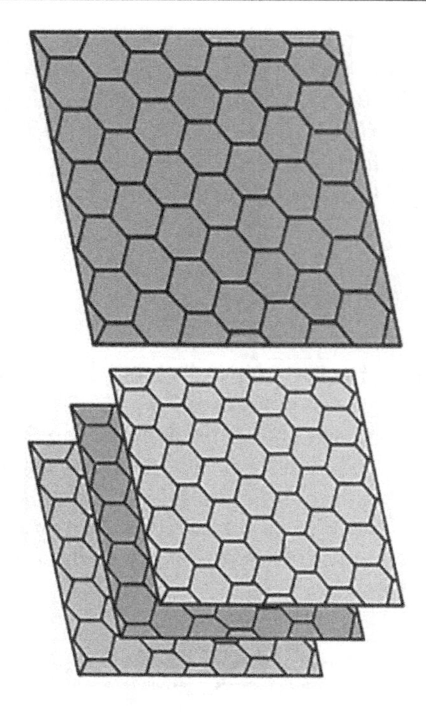

Figure 5.12 The structure of graphene as (a) an SLGNR and (b) an MLGNR.

in 2007 reported that the number of graphene layers are primarily responsible for electronic properties [39]. At the nanoscale regime, understanding the circuit behavior and electronic properties is essential for producing nanoelectronic devices. Recently, GNR as novel material is the primary research area in the production of nanoelectronic devices. Moreover, it's zero band gap also motivates researchers to replace the conventional Cu material due to limited functionality at the nanoscale using metallic-GNR as an interconnect application.

GNRs can be categorized as single- (SLGNR) and multilayered GNR (MLGNR) based on the number of graphene sheets, as depicted in Figures 5.12a and 5.12b, respectively. The SLGNR can be defined only for a single layer of the sheet whereas more than one layer of graphene sheet sandwiched with an interlayer distance of $0.34nm$ is termed an MLGNR. Moreover, it is demonstrated that the armchair CNTs are always metallic in nature, while the chiral indices (n, m) in zigzag CNTs are responsible to have both metallic and semiconducting behavior. However, the statement conflicts in the case of GNR, as the zigzag GNRs are always metallic while armchair GNR can behave as metallic if $N = 3i - 1$ or semiconducting if $N = 3i$ or $3i + 1$ (i is an integer) depending on the number (N) of atoms crosses the width.

5.3.4 Unique Properties of CNTs and GNRs

CNTs and GNRs can be obtained from the carbon allotropes having a unique structure that provides an extraordinary property due to the arrangement of carbon atoms in sp^2 hybridization. It has been reported by several research groups that the CNT and the GNR exhibit an extremely large strength and elasticity due to the tightly packed σ bonding between carbon-to-carbon atoms. The strength of CNT/GNR (up to 48000 $KNmkg^{-1}$) is observed as much higher than carbon steel (154 $KNmKg^{-1}$) and the elastic modulus is approximately 1 TPa or 1000 GPa in the case of SWCNT in comparison to the aluminum and steel that is just 350 GPa and 210 TPa, respectively [38]. The high-strength and -tensile properties have made it possible to easily bend due to an extremely thin monolayer thickness. Apart from this, the good thermal behavior of CNTs and GNRs provides higher transportation of electrons along the tubes, which is also known as ballistic conduction. Experimentally, it is observed that the graphene provides improved thermal conductivity of approximately 3500 $W.m^{-1}K^{-1}$ along the axis as compared to copper at room temperature. The stronger in-plane sigma bonding is primarily responsible for the higher thermal conductivity at temperatures below 20K, and this behavioral change is used to develop nanoelectronic devices for molecular electronics, sensors and actuating devices, flexible electronics, and so on. The CNTs and GNRs also exhibit a unique electrical property based on the position of atoms and chiral indices [38, 39].

5.4 ADVANCED CARBON-BASED HYBRID INTERCONNECT MODELS

The continuous scaling and advancement of technology primarily drive the research towards developing hybrid on-chip interconnect solutions. In this regard, Zhao *et al.* [40] proposed a novel hybrid interconnect structure such as Cu/graphene and graphene/graphene and investigate the performance using the electronic stability control (ESC) circuit model. It is reported that the electrical conductivity can be improved by incorporating graphene as a liner material that can be a suitable choice for interconnect applications. In order to prove the concept, Xu *et al.* [41] also demonstrated the frequency response for Cu–graphene-based interconnect. The partial element equivalent circuit model is used to investigate the performance, and it is observed that the proposed model exhibits improved performance compared to a Cu interconnect. Later, Cheng *et al.* [42] proposed a hybrid interconnect consists of a diffusion barrier layer around the Cu as depicted in Figure 5.13. Based on the geometry depicted in Figure 5.13, a DIL setup is used to derive

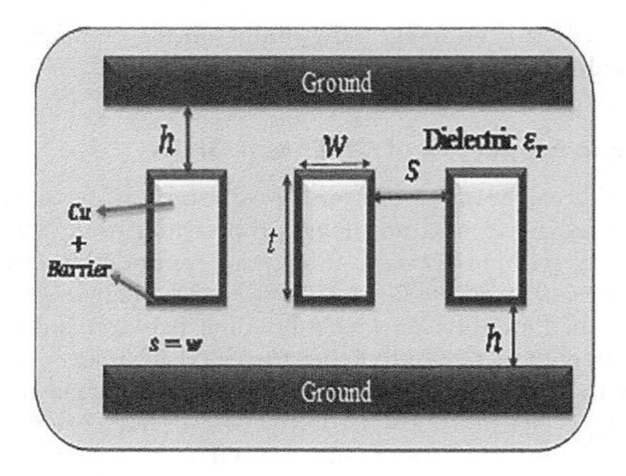

Figure 5.13 Interconnect geometry based on ITRS.

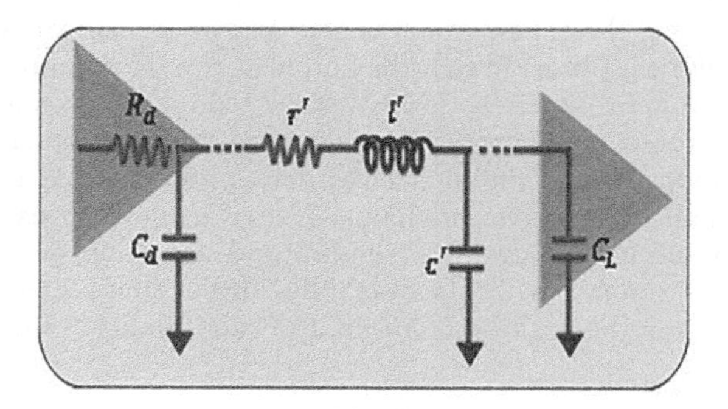

Figure 5.14 An equivalent electrical circuit based on DIL setup.

the equivalent ESC model for hybrid interconnects. In Figure 5.13, the hybrid interconnect is placed on top of the substrate having a distance of h and isolated each interconnect with a spacing of s. The width and thickness of a hybrid interconnect are defined as w and t, respectively, whereas, the effective width, effective thickness and graphene thickness, and the number of layers surrounding the *Cu* of the hybrid interconnect are denoted as $w_{Cu} = w - 2t_g$ and $t_{Cu} = t - 2t_g$, respectively. Depending on the physical parameters, the ESC model consists of equivalent resistance (r'), inductance (l'), and capacitance (c') in *p.u.l* for Cu–graphene interconnect as depicted in Figure 5.14. It is reported that there is a negligible change in the capacitance while an insignificant change in inductance occurs after growing the graphene barrier. The grain size and specularity parameters can be improved using the graphene barrier due to a reduction in effective resistivity. It is also reported that the time-delay performance can be improved by incorporating the graphene barrier instead of the conventional diffusion barriers.

Sun *et al.* [43] developed two distinct models and simulation algorithms to analyze the Cu–graphene-based hybrid interconnect. It is reported that the proposed model exhibits high-frequency operation, and the model's accuracy can be improved by incorporating the dynamic coupling between the graphene. Recently, Kumar *et al.* [44] demonstrated Cu–graphene hybrid on-chip interconnect model to investigate the reliability and signal integrity using temperature and dielectric roughness aware matrix rational approximation (MRA) model. The researchers reported less than 5% deviation in the proposed model compared to the conventional model that can be considered for next-generation interconnect applications.

In order to stop the diffusion of Cu ions into the silicon/dielectric region, a barrier layer, which is sandwiched between Cu lines and dielectric material has been employed. As we move into the sub-22*nm* technology nodes, this barrier layer needs to occupy minimum thickness so that the effective area available for current conduction is not affected. At the same time, the barrier layer should be conducting and should contribute to minimizing the effective resistance of Cu lines. Ta/W/Ti materials are commonly used for barrier layers to prevent the diffusion of Cu ions, and it also resulted in a substantial improvement in the thermal, electrical, and mechanical properties of interconnects [44, 45]. Cu at sub-22*nm* technology nodes exhibit a current density of 10^6 A/cm^2 that has forced designers to look for alternative materials that can meet the foreseeable challenges in ultra-scaled interconnects [45]. To achieve the superior performance of Cu, hybrid interconnects with MLGNRs capping layers are seen as a favorable interconnect technology for the future of technology nodes. A comparison of different barrier layer materials is given in Table 5.2 [45].

Advantages of graphene capping on Cu material results in improvement of breakdown current density by 18% [46] and an increase in the thermal conductivity by 27% [47]. Goli *et al.* [47] depict that for micrometer dimensions, graphene deposition on Cu reduces the surface and grain boundary scattering by significantly enlarging the grain sizes when compared to that

Table 5.2 Resistivity of Tungsten, MLGNR, and Tantalum Barrier Layers

Thickness Resistivity of Barrier layer (Ω-m)	*2nm*	*1.2nm*	*0.6nm*
Tungsten	65.1×10^{-8}	100×10^{-8}	194.2×10^{-8}
Tantalum	278.6×10^{-8}	433.2×10^{-8}	852.4×10^{-8}
MLGNR	8×10^{-8}	10.22×10^{-8}	14×10^{-8}

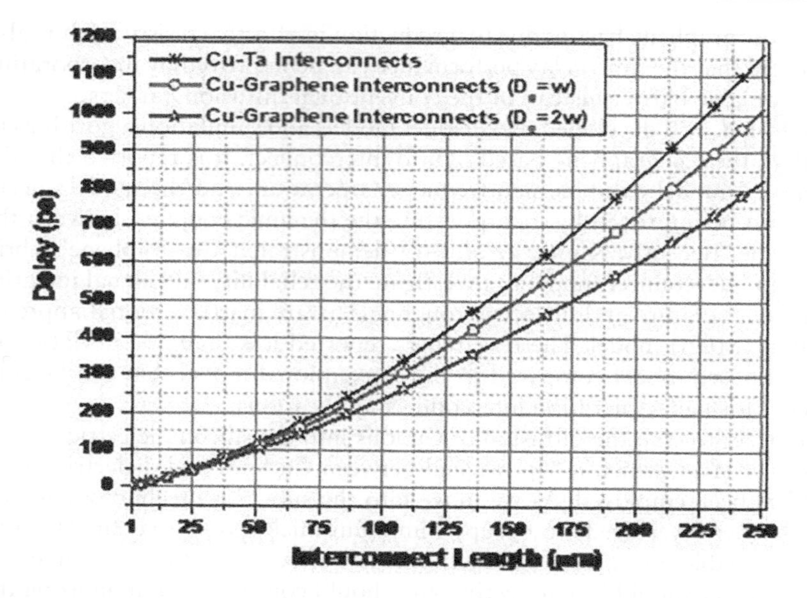

Figure 5.15 Comparison of the propagation delay of Cu interconnects at the 13*nm* technology node with a different barrier layer as a function of the interconnect length [18].

of conventional Cu material. Also, the surface roughness reduces significantly. Mehta *et al.* [48] have reported that graphene as a barrier layer on Cu surfaces results in nearly 15% faster data rates as compared to that in Cu interconnects. The deposition of graphene layers on Cu increases its activation energy [49], preventing the diffusion of Cu into the dielectrics and reducing any scattering by the enlargement in the grain sizes. Also, multilayer graphene capping on Cu exhibits longer electromigration time and mean time to failure (MTTF) as compared to Cu interconnects. The graphene layer on the Cu is also believed to improve the current carrying capacity beyond the ITRS projected roadmap for Cu. Figure 5.15 shows a significant reduction in the resistivity of Cu–graphene interconnects as compared to Cu–TaN interconnects, and its impact on resistivity becomes more significant by considering an enlarged grain size [45].

It is observed that the Cu–graphene interconnects outperformed the Cu–TaN interconnects. The propagation delay of Cu–graphene hybrid interconnects is significantly improved, when the morphological change in Cu after the growth of graphene over it is considered. As the length of interconnects increases, the propagation delay of Cu–graphene interconnects keeps improving. Although there is rapid progress in the fabrication process of CNTs and GNRs, the current nanoscale technology still faces the integration challenges associated with the novel carbon-based material. Contact resistance, diameter control, CNT alignment, chirality problem, and GNR edge roughness are some of the issues related to the carbon-based interconnects.

5.5 FUTURE TECHNOLOGIES AND TRENDS

Below the *7nm* technology node, interconnects exhibit major changes in their physical and electrical properties. The maximum effect of scaling is seen on the first or second layers of interconnects. As scaling continues, the barrier layers also need to scale without any loss of conductivity and reliability concern becomes important when the barrier layer reaches *1nm* or *2nm*. Even if one atom is missing, the barrier layer ceases to exist [50]. Therefore, below *7nm*, we need to get ready for newer interconnect structures, materials, and novel computational technologies, that is, artificial neural networks, machine learning, and the like [51].

Below the *10nm* technology node, current density increases dramatically and results in greater Joule heating. Thermal effects are an inseparable aspect of interconnects due to self-heating caused by the flow of current and due to environmental heating in high-speed designs. The grain and surface scattering increase with temperature, and resistivity further increases with scaling, causing severe issues in nanoscale interconnects [44]. By increasing the temperature, the MFP of the interconnects decreases, and consequently, the interconnect resistivity increases. Also, the crosstalk delay rises with an increase in the temperature. Effective resistivity in hybrid interconnects is lesser than that in conventional Cu interconnects, particularly at lower technology nodes and higher temperatures. Similarly, it can be seen observed that the crosstalk delay of MLGNR increases at higher temperatures due to increased resistance. In three-dimensional (3D) stacked ICs, thermal management is one of the most challenging problems. By increasing the number of layers, the heat dissipation per unit volume and footprint area is significantly increased. Also, the inner layers in 3D ICs are not directly connected to the heat sink, which leads to significant heating issues [52]. So, it is pretty important to determine the electrical, physical, and thermal sensitivity of the interconnect links through electrical-thermal codesign methodology.

5.5.1 Electronic-Optic Integration

Looking at the future, one can imagine a world of faster data transfer and quantum computing. The integration of two technologies, that is, electronics and optics layers as one unit on the same chip, requires accurate codesign through novel fabrication processes. The key applications of this integration lay in various quantum technologies and biosensors, to name a few. The future demand in this field depends on the ability to connect both these

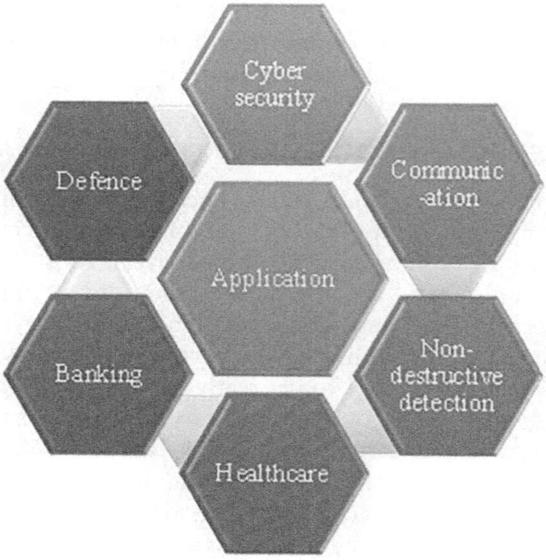

Figure 5.16 Application areas for electronic-photonic systems.

disparate technologies at the nanoscale dimensions. Different applications based on this hybrid electro-optic integration are listed in Figure 5.16. The integration of optics and electronics will form an electronic-optic technology that can be exploited in various ways:

(a) Ultra-fast data transfer between memory and processor
(b) High-fidelity optical-signal processing using communication chips
(c) Highly parallel biochemical sensors

However, the semiconductor manufacturing industry lacks semiconductor materials with appropriate optical properties for realizing these active and passive photonic functions. Therefore, all attempts of integrating photonics and electronics into CMOSs have been limited to silicon-on-insulator substrate. Apart from this, there is a limited supply chain that has kept the manufacturing costs pretty high.

5.5.2 Machine Learning for Analysis of High-Speed Interconnects

Modeling and performance analysis of on-chip interconnect structures are primarily done through signal integrity analysis that involves time-domain analyses of understanding the eye diagram and computing bit error rates.

However, bit rates increase with every new technology node. This makes it quite difficult to perform time-domain analysis due to the highly computationally intensive and time-consuming process. A machine learning-based approach can address these challenges in a far more efficient manner.

5.6 SUMMARY

The need for scaling in the semiconductor industry restricts the use of Cu as an interconnects material at lower technology nodes. At high frequencies and lower technology nodes, Cu interconnects are suffered with size effect, electromigration, higher coupling noise, reliability issues, and dispersion effects. GNRs and CNTs can be seen as potential materials to replace Cu for interconnect applications to mitigate all these issues. However, GNRs and CNTs higher growth temperature and fabrications challenges may limit the performance of such emerging materials. Therefore, Cu–graphene hybrid interconnects can commendably fulfill these future requirements. Moreover, the use of novel techniques such as graphene-based through-silicon vias, ThruChip Interfaces, spintronics, plasmonic and optical-based interconnects can be envisaged toward the design and fabrication of ultra-high-performance heterogeneous ICs.

REFERENCES

[1] Davis, J.A., et al., "Interconnect Limits on Gigascale Integration (GSI) in the 21st Century", Proceedings of the IEEE, Vol. 89, No. 3, PP. 305–324, 2001.

[2] Gardner, D.S., et al., "Layered and Homogeneous Films of Aluminum and Aluminum/Silicon with Titanium and Tungsten for Multilevel Interconnects", IEEE Journal of Solid-State Circuits, Vol. 20, No. 1, PP. 94–103, 1985.

[3] Kapur, P., G. Chandra, J.P. McVittie, and K.C. Saraswat, "Technology and Reliability Constrained Future Copper Interconnects: Performance Implications", IEEE Transactions on Electron Devices, Vol. 49, No. 4, PP. 598–604, 2002.

[4] Travaly, Y., et al., "On a More Accurate Assessment of Scaled Copper/low-k Interconnects Performance", IEEE Transactions on Semiconductor Manufacturing, Vol. 20, No. 3, PP. 333–340, 2007.

[5] Zhao, L., "All About Interconnects", December 2017. https://semiengineering.com/all-about-interconnects/

[6] Havemann, R.H., and J.A. Hutchby, "High-performance Interconnects: An Integration Overview", Proceedings of the IEEE, Vol. 89, No. 5, PP. 586–601, 2001.

[7] Kumar, S., and R. Sharma, "Analytical Modeling and Performance Benchmarking of On-chip Interconnects with Rough Surfaces", IEEE Transactions on Multi-Scale Computing Systems, Vol. 4, No. 3, PP. 272–284, 2018.

[8] Kumar, S., and R. Sharma, "Chip-to-chip Copper Interconnects with Rough Surfaces: Analytical Models for Parameter Extraction and Performance Evaluation", IEEE Transactions on Components, Packaging and Manufacturing Technology, Vol. 8, No. 2, PP. 286–299, 2018.

[9] Banerjee, K., and A. Mehrotra, "A Power-optimal Repeater Insertion Methodology for Global Interconnects in Nanometer Designs", IEEE Transactions on Electron Devices, Vol. 49, PP. 2001–2007, 2002.
[10] Cheng, Y., C. Lee, and Y. Huang, "Copper Metal for Semiconductor Interconnects", in Noble and Precious Metals—Properties, Nanoscale Effects and Applications. IntechOpen, United Kingdom, 2018.
[11] Jeffery, G., "Process Technology for Copper Interconnects", in Handbook of Thin Film Deposition, 3rd ed., Elsevier, Netherlands, 2012, PP. 221–269.
[12] Kaushik, B.K., M.K. Majumder, and V.R. Kumar, "Carbon Nanotube Based 3-D Interconnects—A Reality or a Distant Dream", IEEE Circuits and Systems Magazine, Vol. 14, No. 4, PP. 16–35, 2014.
[13] Im, S., et al., "Scaling Analysis of Multilevel Interconnect Temperatures for High-Performance ICs", IEEE Transactions on Electron Devices, Vol. 52, No. 12, PP. 2710–2719, 2005.
[14] Srivastava, N., et al., "A Comparative Scaling Analysis of Metallic and Carbon Nanotube Interconnections for Nanometer Scale VLSI Technologies", in Proceedings of the 21st International VLSI Multilevel Interconnect Conference (VMIC), Sept. 29–Oct. 2, Waikoloa, HI, pp. 393–398, 2004.
[15] Biagalke, S., et al., "Load-aware Redundant Via Insertion for Electromigration Avoidance", in Proceedings of ACM International Symposium on Physical Design, 2016, PP. 99–106. https://doi.org/10.1145/2872334.2872355.
[16] Chen, D., E. Li, E. Rosenbaum, and S. Kang, "Interconnect Thermal Modeling for Accurate Simulation of Circuit Timing and Reliability", IEEE Transactions on Computer-Aided Design of Integrated Circuits and Systems, Vol. 19, No. 2, PP. 197–205, 2000.
[17] Steinhogl, W., G. Schindler, G. Steinlesberger, M. Traving, and M. Engelhardt, "Comprehensive Study of the Resistivity of Copper Wires with Lateral Dimensions of 100 nm and Smaller", Journal of Applied Physics, Vol. 97, No. 2, PP. 023 706-1–023 706-7, 2005.
[18] Kumar, S., and R. Sharma, "Analytical Model for Resistivity and Mean Free Path in On-chip Interconnects with Rough Surfaces", IEEE Transactions on Emerging Topics in Computing, Vol. 6, No. 2, PP. 233–243, 2018.
[19] Mistry, K., "The Forever Exponential? Moore's Law: Past, Present and Future", in Proceedings of IEEE Electron Devices Technology & Manufacturing Conference (EDTM), Penang, Malaysia, 2020, PP. 1–1.
[20] Dennard, R.H., et al., "Design of Ion-implanted MOSFET's with Very Small Physical Dimensions", IEEE Solid-State Circuits Society Newsletter, Vol. 9, No. 5, PP. 38–50, 2007.
[21] Kumar, S., and R. Sharma, "Design of Energy-aware Interconnects for Next Generation Micro Systems", CSI Transactions on ICT, Springer, Vol. 7, No. 3, PP. 215–220, 2019.
[22] International Technology Working Groups, "International Technology Roadmap for Semiconductors (ITRS)", 2013. http://www.itrs2.net/2013-itrs.html.
[23] Balamurugan, G., et al., "A Scalable 5–15 Gbps, 14–75 mW Low-power I/O Transceiver in 65 nm CMOS", IEEE Journal of Solid-State Circuits, Vol. 43, No. 4, PP. 1010–1019, 2008.
[24] Kumar, V., et al., "Compact Modeling and Performance Optimization of 3D Chip-to-chip Interconnects with Transmission Lines, Vias and Discontinuities",

in Proceedings ofIEEE International Interconnect Technology Conference, USA, 2012, PP. 1–3.
[25] Hou, Y., and C.M. Tan, "Size Effect in Cu nano-interconnects and Its Implication on Electromigration", in Proceedings ofIEEE International Nanoelectronics Conference, Shanghai, 2008, PP. 610–613.
[26] Sondheimer, E.H., "The Mean Free Path of Electrons in Metals", Advances in Physics, Vol. 50, No. 6, PP. 499–537, 2001.
[27] Fuchs, K., "The Conductivity of Thin Metallic Films According to the Electron Theory of Metals", Mathematical Proceedings of the Cambridge Philosophical Society, Vol. 34, No. 1, PP. 100–108, 1938.
[28] Kurusu, T., H. Tanimoto, M. Wada, A. Isobayashi, A. Kajita, N. Aoki, et al., "A Monte Carlo Simulation of Electron Transport in Cu nano-interconnects: Suppression of Resistance Degradation Due to LER/LWR", in Proceedings of IEEE International Electron Devices Meeting, USA, 2012, PP. 30.7.1–30.7.4.
[29] Kumar, S., et al., "Crosstalk Analysis for Rough Copper Interconnects Considering Ternary Logic", in Proceedings of IEEE Electrical Design of Advanced Packaging and System Symposium (EDAPS), Chandigarh, India, 2018, PP. 1–3.
[30] Lopez, G., J. Davis, and J. Meindl, "A New Physical Model and Experimental Measurements of Copper Interconnect Resistivity Considering Size Effects and Line-edge Roughness (LER)", in Proceedings of IEEE International Interconnect Technology Conference, Japan, 2009, PP. 231–234.
[31] Kang, C.G., et al., "Effects of Multi-layer Graphene Capping on Cu Interconnects", Nanotechnology, Vol. 24, No. 11, P. 115707, 22 March 2013.
[32] Lienig, J., and M. Thiele, Fundamentals of Electromigration-Aware Integrated Circuit Design. Springer Publishing Company, Incorporated, Germany, 2018.
[33] Filippi, R.G., et al., "The Effect of a Threshold Failure Time and Bimodal Behavior on the Electromigration Lifetime of Copper Interconnects", Proceedings of 2009 IEEE International Reliability Physics Symposium, Montreal, QC, Canada, 2009, PP. 444–451.
[34] https://www.wsts.org/76/Recent-News-Release.
[35] Kumbhare, V.R., P.P. Paltani, and M.K. Majumder, "Impact of Interconnect Spacing on Crosstalk for Multi-layered Graphene Nanoribbon", IETE Journal of Research, PP. 1–10, 2019.
[36] Taniguchi, N., "On the Basic Concept of 'Nano-Technology'", in Proceedings of the International Conference on Production Engineering Tokyo, Part II, Japan Society of Precision Engineering, Tokyo, 1974, PP. 5–10.
[37] Iijima, S., and T. Ichihashi, "Single-shell Carbon Nanotubes of 1-nm Diameter", Nature, Vol. 363, No. 6430, PP. 603–605, 1993.
[38] Majumder, M.K., V.R. Kumbhare, A. Japa, and B.K. Kaushik, Introduction to Microelectronics to Nanoelectronics: Design and Technology, 1st ed., CRC Press, Taylor & Francis, FL, 2020.
[39] Wakabayashi, K., M. Fujita, H. Ajiki, and M. Sigrist, "Electronic and Magnetic Properties of Nanographite Ribbons", Physical Review B, Vol. 59, No. 12, PP. 8271–8282, 1999.
[40] Zhao, W.S., D.W. Wang, G. Wang, and W.Y. Yin, "Electrical Modeling of On-chip Cu-graphene Heterogeneous Interconnects", IEEE Electron Device Letters, Vol. 36, No. 1, PP. 74–76, 2015.

[41] Xu, Y., Y.S. Li, D. Yi, X. Wei, and E. Li, "Signal Transmission along Cu-graphene Heterogeneous Interconnects", in Proceedings of 2016 Asia-Pacific International Symposium on Electromagnetic Compatibility (APEMC), Shenzhen, 2016, PP. 1007–1009.
[42] Cheng, Z., et al., "Analysis of Cu-graphene Interconnects", IEEE Access, Vol. 6, PP. 53499–53508, 2018.
[43] Sun, S., and D. Jiao, "First-principles-based Multiphysics Modeling and Simulation of On-chip Cu-graphene Hybrid Nanointerconnects in Comparison with Simplified Model-based Analysis", IEEE Journal on Multiscale and Multiphysics Computational Techniques, Vol. 4, PP. 374–382, 2019.
[44] Kumar, R., et al., "A Temperature and Dielectric Roughness-aware Matrix Rational Approximation Model for the Reliability Assessment of Copper—graphene Hybrid On-chip Interconnects", IEEE Transactions on Components, Packaging and Manufacturing Technology, Vol. 10, No. 9, PP. 1454–1465, 2020.
[45] Kumar, R., et al., "Role of Grain Size on the Effective Resistivity of Cu-graphene Hybrid Interconnects", in Proceedings of IEEE Electronic Components and Technology Conference (ECTC), Orlando, FL, 2020, PP. 1620–1625.
[46] Li, L., Z. Zhu, A. Yoon, and H.P. Wong, "In-situ Grown Graphene Enabled Copper Interconnects with Improved Electromigration Reliability", IEEE Electron Device Letters, Vol. 40, No. 5, PP. 815–817, 2019.
[47] Goli, P., H. Ning, X. Li, C.Y. Lu, K.S. Novoselov, and A.A. Balandin, "Thermal Properties of Graphene–copper–graphene Heterogeneous Films", Nano Letters, Vol. 14, No. 3, PP. 1497–1503, 2014.
[48] Mehta, R., S. Chugh, and Z. Chen, "Enhanced Electrical and Thermal Conduction in Graphene-encapsulated Copper Nanowires", Nano Letters, Vol. 15, No. 3, PP. 2024–2030, 2015.
[49] Yoon, S.J., A. Yoon, W.S. Hwang, S. Choi, and B.J. Cho, "Improved Electromigration-resistance of Cu interconnects by Graphene-based Capping Layer", in Proceedings of IEEE Symposium on VLSI Technology (VLSI Technology), Japan, 2015, PP. T124–T125.
[50] Sperling, E., "Big Changes in Tiny Interconnects", 2020. https://semiengineering.com/big-changes-in-tiny-interconnects/
[51] Kasai, R., T. Kanamoto, M. Imai, A. Kurokawa, and K. Hachiya, "Neural Network-based 3D IC Interconnect Capacitance Extraction", in Proceedings of IEEE International Conference on Communication Engineering and Technology (ICCET), Nagoya, Japan, 2019, PP. 168–172.
[52] Yip, T.G., W.T. Beyene, G. Kollipara, W. Ng, and J. Feng, "Electrical-thermal co-Design of High Speed Links", in Proceedings of IEEE Electronic Components and Technology Conference (ECTC), Las Vegas, NV, 2010, PP. 1893–1899.

Chapter 6

Performance Review of Static Memory Cells Based on CMOS, FinFET, CNTFET and GNRFET Design

G. Boopathi Raja

CONTENTS

6.1 INTRODUCTION

Future nanotechnology can result in a doubling of transistor size and miniaturization of transistors in electronic devices. Currently, we have the Intel Core i7 processor; it consists of a single integrated circuit (IC), which has

DOI: 10.1201/9781003311379-6

more than 600 million transistors. As a result, scaling and miniaturization affect nano-transistor performance. Because of short channel effects, such as subthreshold leakage current, downscaling the transistor has become a challenging and difficult task.

To achieve ultra-high precision, more than Moore's law inventions and new nanostructures must be added. A new architecture, modified design, and the use of a certain material are all examples of these new methods [1]. The greater the surface areas of the SoC (system on chip), the more transistors are included in the circuit design.

Since the dimensions of the gate region of the field-effect transistor (FET) have been reduced, the number of SRAM cells in the memory chip can be increased. The conventional planar metal–oxide–semiconductor field-effect transistor (MOSFET), on the other hand, suffers from several problems. The most important issues are the threshold voltage (Vth) problem and short channel effects as the technology scales beyond 32 nm. The 6T SRAM cells based on 16-nm fin field-effect transistors (FinFETs) may be a lot more efficient than planar MOSFETs.

Graphene nanoribbon field-effect transistors (GNRFETs) and carbon nanotube field-effect transistors (CNTFETs) are two examples of carbon-based materials that can boost system performance, not only in terms of reliability but also in terms of power consumption. The performances of carbon-based devices are more efficient when compared with FinFET technology in terms of speed.

Strained silicon, gallium arsenide (GaAs) and high-K dielectric materials will improve the performance of a device, resulting in considerable gate power, additionally based on their ability to minimize short-channel effects. The main objective of graphene nanoribbon (GNR)–based FET preferred in logic circuits to resolve the limitations of traditional planar MOSFETs and their efficiency as static memory cells is investigated [1].

The advancement of nanoelectronics-based semiconductor devices has resulted in the nanometer scaling of transistor channels. However, several issues arise when downscaling transistors, the most common of which are numerous short channel effects.

The problems associated with conventional planar MOSFETs due to quantum mechanical (QM) effects were solved by GNRFET. Due to its excellent electrical properties, the graphene nanoribbon–based transistor may be preferred in static memory cell circuit architecture. When used as a high-speed memory cache, static memory cells are more powerful and quicker. The standard 6T SRAM cell suffers from transistor-size constraints as an export between access and write stability.

A few recent studies compared the performance of 15-nm GNRFET-based 6T and 8T SRAM cells to 16-nm MOSFET and 16-nm FinFET cells in terms of access, retention, and write. The SRAM model was created and simulated using Synopsys HSPICE. Based on power consumption and static noise margin (SNM), FinFET-, GNRFET-, and MOSFET-based 8TSRAM cells outperform six transistor SRAM cells. Compared to FinFET, GNR-FET, and MOSFET-based 6T SRAM configurations, the FinFET, GNRFET, and MOSFET-based 8T SRAM cells improved access static noise margin of approximately 60 percent, 30 percent, and 21 percent, respectively, as well as total power usage about 97 percent, 99 percent, and 83 percent, respectively, [1].

The chapter is organized as follows: Section 6.2 describes the outline of complementary MOSFET (CMOS)–based logic circuit design. In this section, the advantages and disadvantages of CMOS technologies were discussed along with possible alternate devices. The prominent alternatives to CMOS devices such as FinFET, CNTFET, GNRFET, and others, are discussed in Section 6.3. The conventional 6T SRAM cell and other modified forms were listed in Section 6.4. In Section 6.5, the obtained simulation results were discussed, and finally, our conclusions are discussed in Section 6.6.

6.1.1 Limitations of Moore's Law

Moore's law states that the transistors used in an integrated circuit may double per year. It has held for a very long time [2].

Moore's law is focused on transistors shrinking, and physics will inevitably interfere. This is difficult for chip designers. Due to electron tunneling, the length of a gate—the portion of a transistor that controls the movement of electrons on or off—cannot be less than 5 nm.

The limitations of Moore's law have been overtaken by Neven's law. Hartmut Neven, the director of Google's Quantum AI Lab, is the inspiration for Neven's law.

6.2 CMOS-BASED CIRCUIT DESIGN

In today's CMOS very large-scale integration (VLSI) circuit architecture, power dissipation has become a major issue. In the case of battery life in battery-operated systems, maximum power dissipation is not considered as desirable because it affects aspects like reliability and ventilation costs and leads to battery life reduction [3–5].

One of the most critical problems in today's CMOS VLSI architecture is leakage and power dissipation. A leakage-free CMOS circuit design is a difficult challenge. There are a few strategies for reducing the power dissipation of CMOS VLSI circuits that function well. In this approach, the stack ONOFF IC technique was preferred for minimizing leakage power in CMOS-based logic circuits.

6.2.1 Advantages of CMOS Technology

There are a few benefits of using CMOS logic [6]:

- High Input Impedance—a layer of insulation separates the electrodes from the material they are controlling, and the input signal drives them (metal oxide). As a result, they have a low capacitance but a nearly infinite resistance. Leakage occurs when the current into or out of a CMOS input is retained at a single stage, which is usually 1nA or less.
- Both directions are actively driven by the outputs.
- The outputs are almost rail to rail, and when kept fixed, CMOS logic consumes very little power. As capacitors are charged and discharged, switching induces current consumption. Also, it has a better speed-to-power ratio on comparing with other logic models.

The design of CMOS based logic circuit was quite simple. An inverter, which consists of just two transistors, is the simplest gate. This, along with its low power consumption, makes it ideal for dense integration. Alternatively, a lot of justification can be seen for the cost, area, and power.

6.2.2 Disadvantages of CMOS Technology

However, CMOS technology has suffered from a lot of problems [6–8]:

- It is not bipolar.
- Certain circuits are not feasible.
- Execution is difficult.

The following are the drawbacks of using CMOS technologies to apply vision chips:

- **Analog circuit design:** For analog circuit design, leading-edge processes are not categorized and optimized.
- **Photodetectors:** In very few of the processes are the photodetector mechanisms characterized. The designer must ensure that the photodetectors perform as anticipated.
- **Second-order effects:** Some second-order interface properties, such as subthreshold operation, are typically ignored or given less weight throughout the scaling process, and their elimination is preferred over their augmentation.
- **Mismatch:** In CMOS devices, the mismatch is comparatively large. This has a particularly negative impact on the dependability of analogue processing in vision chips [9–12].

6.2.3 Alternatives to CMOS Technology

New integrated circuit technology was developed by the semiconductor industry that will take us beyond the CMOS age, allowing computing to continue to advance in terms of power and efficiency. As an alternative to CMOS technology, researchers are currently investigating new system concepts and new knowledge keys.

Quantum electronic devices, including the tunneling field-effect transistor (TFET), as well as devices based on electron spin and spintronics, as well as nano-magnetics, are active research areas. Decisions must be taken in the next ten years to find suitable CMOS alternatives by 2025 [13–16].

Benchmarking methods and measurements are being used to refine and direct research discovery in materials, computers, and circuits. This discussion includes an outline of the research horizon beyond CMOS applications, as well as benchmarking of these devices for computation.

6.3 ALTERNATIVE DEVICES

When it comes to scaling down the dimensions of CMOS transistors, there has been a slew of issues. Because of the excellent electrical properties and integration capabilities offered by improved design techniques, transistors built from carbon-based nanomaterials have emerged as exciting next-generation products [17–20]. Carbon nanotube (CNT) transistors and GNR-based transistors are the most studied FETs. GNRs may be cultivated in situ using a silicon-compatible, transfer-free process, unlike cylindrical CNTs [19, 21, 22], with no compatibility or transfer-related issues like CNT-based circuits [23]. Due to process variation, graphene-based circuits have several issues including reduced mobility, narrow band gap, and unreliable conductivity [24–27].

6.3.1 FinFETs

FinFET is a kind of nonplanar transistor, sometimes known as a three-dimensional transistor. It is the foundation for the creation of contemporary nanoelectronic semiconductor devices. FinFET gates were originally introduced in the early 2010s, and they soon became the most common gate design at manufacturing nodes of 14 nm, 10 nm, and 7 nm. A single FinFET transistor often has multiple fins that are positioned side by side and all covered by the same gate to improve driving strength and performance.

FinFETs were implemented to satisfy the specifications of the semiconductor road map when existing systems reached their scaling limits, and they were established as one of the alternatives to CMOS-based devices. It will most likely be made with a typical CMOS technique, while an ideal silicon-nanowire while having superior performance, is difficult to fabricate [28].

6.3.2 Nanowires

Nanowires are nanostructures having a diameter of less than one nanometer (10^9 meters). It may alternatively be described as a length-to-width ratio that is larger than 1000. Nanowires can also be described as structures with a thickness or width of tens of nanometers or less, as well as an unrestricted length.

FinFETs may be introduced as a way to fulfill the semiconductor roadmap's expectations. While an optimal silicon nanowire performs excellently, it provides a challenging task. According to the observations, the FinFET structure could be the best architecture for electrical properties that are comparable to silicon nanowires [28–30].

6.3.3 CNTFETs

In ternary logic circuits based on the standard CMOS, multiple thresholds are needed. MOS transistors can be body-biased to achieve this. This makes circuit design more difficult and demands the use of alternate components such as CNTFETs.

Researchers are investigating various post-silicon, post-binary logic technologies in light of the issues that have arisen as a result of the scaling of the silicon transistor. To implement ternary logic circuits, one choice is to use CNTFETs. By changing the physical size of CNTFETs, a wide range of threshold voltages may be obtained, making them an attractive choice for creating ternary logic circuits.

By changing the physical size of the CNTs, multiple threshold voltages in CNTFETs may be obtained, making them a superior alternative for ternary logic circuits. Chirality is a CNTFET feature that changes the threshold voltage based on the CNT diameter. CNTFETs with various chiralities can be used to reach a variety of thresholds needed for the implementation of ternary logic circuits. CNTFET technology can replace CMOS technology because it can create circuits that are up to ten times more energy-efficient.

6.3.4 GNRFETs

GNRFET is a remarkable advancement in the world of electronics that has gained considerable attention. According to recent simulations of GNRFET circuits, GNRFETs may be useful in low-power applications.

MOS-GNRFET utilizes 18 percent and 54 percent of total power, respectively, compared to high-performance (HP) Si-CMOS and low-power (LP) MOS-GNRFET. SB-GNRFET is not as efficient as MOS-GNRFET when it comes to power consumption. The optimum (non-ideal) SB-GNRFET has a 3 percent (5.4×) and 0.45 percent (83.5%) energy-delay product (EDP), respectively, compared to Si-CMOS (HP) and Si-CMOS (LP).

6.4 STATIC MEMORY CELLS

RAM is a type of memory that permits users to read and update data and machine code in any order. It is frequently used to keep track of working data and machine code. Unlike other direct-access data storage media, where mechanical constraints cause the time, it takes to read and write data items to vary considerably based on their actual placement on the recording medium. Regardless of their actual position inside the memory, a RAM device allows data objects to be read or written in approximately the same amount of time. To connect the data lines to the chosen storage for reading or writing the entry, RAM includes multiplexing and demultiplexing hardware. RAM devices generally contain multiple data lines and are labeled as 8-bit, 16-bit, and so on because the same address frequently accesses more than one bit of storage [31–33].

MOS memory cells are currently used to create random-access memory (RAM) on integrated circuit (IC) devices. Dynamic random-access memory (DRAM) and static random-access memory (SRAM) are the two most common forms of volatile random-access semiconductor memory (SRAM). Commercial MOS memory, which makes use of MOS transistors, was invented in the late 1960s and has since become the foundation for all commercial semiconductor memory [34–36].

Despite the implementation of non-volatile RAM, RAM is most often used in conjunction with volatile memory, such as DRAM modules, which lose data when power is removed. Several forms of nonvolatile memory offer random access for read operations but not for writes or have other constraints. Most varieties of ROM and NOR-Flash flash memory fall within this group. Static memory is a form of digital memory that preserves data even when the power is turned off.

6.4.1 Conventional 6T SRAM Cells

Back-to-back connected inverters were used to construct the traditional 6T SRAM memory cell. A p-channel MOS transistor (PMOS) and an n-channel MOS transistor are required for each inverter cell (NMOS). To access the data, two additional NMOS transistors were needed. A bit line is connected to one access transistor, and a bit line bar is connected to another. Figure 6.1 shows the circuit diagram of a standard 6T SRAM cell.

The memory cell performs the operation in anyone of the following modes:

1. Read mode
2. Write mode
3. Hold mode (storage mode)

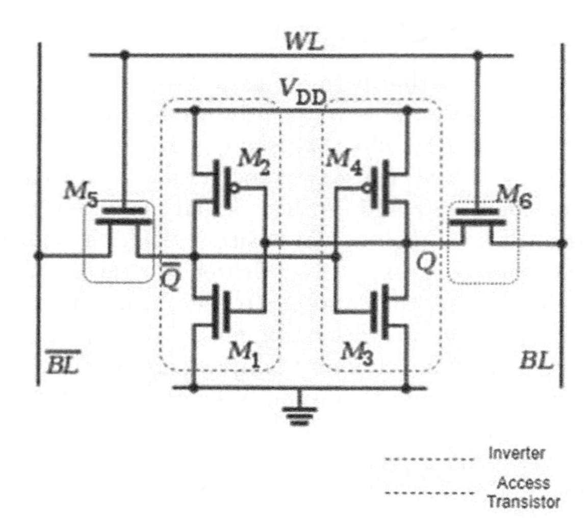

Figure 6.1 Circuit diagram of a standard 6T SRAM cell.

6.5 MODIFIED SRAM CELLS

The performance of conventional 6T SRAM cells have suffered from a lot of problems. This may be overcome by modified design and/or the use of alternate devices instead of CMOS transistors. The drawbacks of conventional 6T SRAM cell are discussed in the following sections.

6.5.1 5T SRAM Cells

The CMOS 5T SRAM cell was chosen for applications requiring high density and LP. The circuit diagram of 5-transistor static memory cell is shown in the Figure 6.2. According to leakage current and constructive feedback, this cell saves data without requiring a refresh cycle. One word line, one bit line, and an external read-line power supply are included in this 5T SRAM cell [37–39].

6.5.2 7T SRAM Cells

The read and write stability of the modified 7T SRAM cell is better than that of ordinary 6T SRAM cells. The access transistor size issue in this cell is resolved by separating read and write access transistors. As a result, big write-access transistors and tiny read-access transistors are used, resulting in improved read and write stability [40–42].

Furthermore, by separating the storing node from the read path, read stability is improved even further. The number of write drivers in this cell is

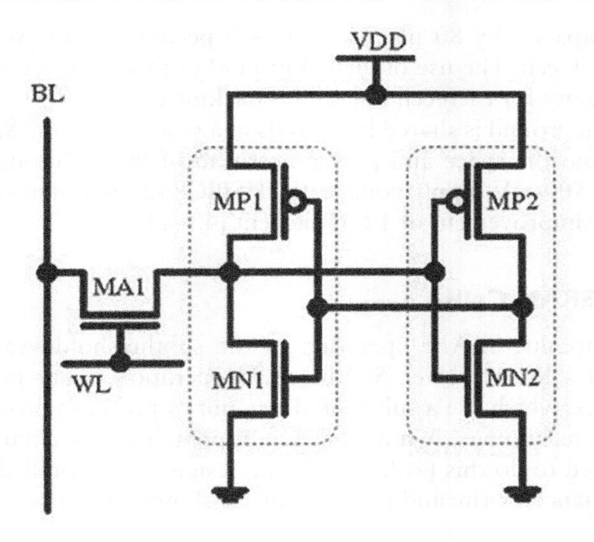

Figure 6.2 Circuit diagram of a standard 5T SRAM cell.

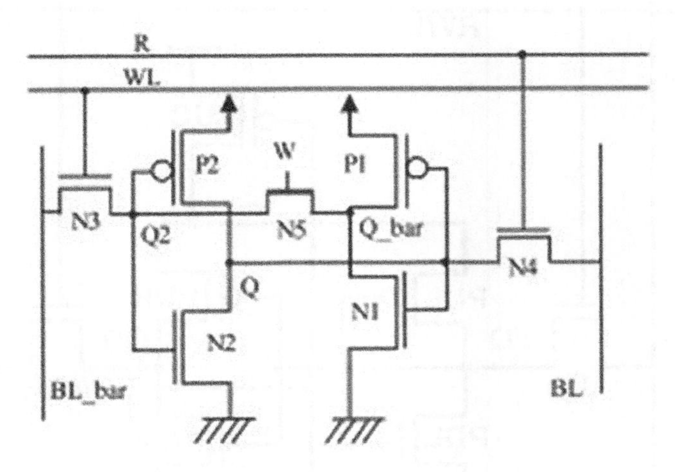

Figure 6.3 Circuit diagram of a 7T SRAM cell.

restricted by a single-ended write process. Virtual ground is used with one of the inverters to increase the writing potential much more. This method decreases positive feedback and increases the cell's writing capacity.

Figure 6.3 shows the circuit diagram of low-power seven-transistor static RAM cell. At a supply voltage of 500 mV, HSPICE simulation in 90-nm CMOS technology proves that the designed structure increases read stability and write capacity by 80 percent and 54.9 percent, respectively, over the traditional 6T cell. The use of virtual ground in this architecture decreases leakage capacity for each cell due to the stacking effect [40].

The virtual ground is shared by all cells in a row to decrease SRAM block regulation and the space and power overhead of the additional transistor required. At 500-mV supply voltage, the HSPICE simulation results reveal a static power improvement of 12.35 percent [43–45].

6.5.3 8T SRAM Cells

At digital circuits, SRAM operation in the subthreshold/weak inversion region saves a lot of power. Subthreshold operation is not possible with SRAM arrays, which use a substantial amount of power in processors with sub-100-nm technology. A novel SRAM approach at the circuit or design level is needed to do this [46]. The circuit design for a modified 8T SRAM cell with enhanced write and read margins is shown in Figure 6.4.

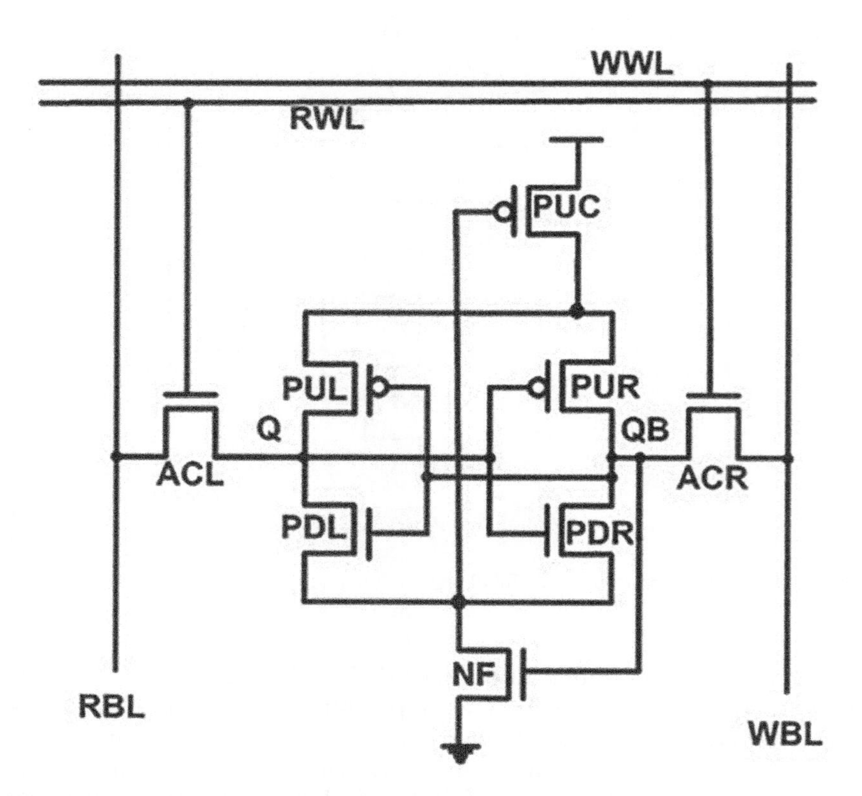

Figure 6.4 Circuit diagram of Modified 8T SRAM cell.

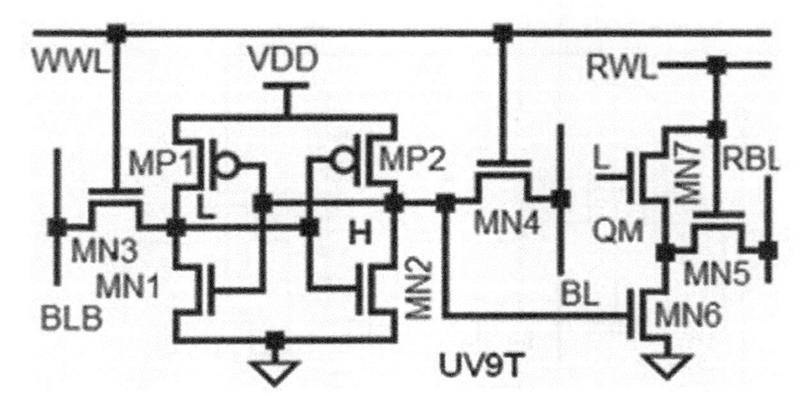

Figure 6.5 Ultra-low-voltage 9-transistor SRAM cell (UV9T).

An 8T-SRAM cell with a supply voltage of 1 V is proposed, exhibiting a significant write margin increase of at least 22 percent over a standard 6T-SRAM cell. In addition, as compared to a standard 6T-SRAM cell, the cell's read static noise margin is enhanced by at least 2.2 times.

Although the SRAM cell increases the overall leakage potential for the super threshold region, it can run at supply voltages as low as 200 mV, reducing total power consumption and increasing the cell's robustness [47, 48].

6.5.4 9T SRAM Cells

Figure 6.5 shows the circuit diagram of UV9T, which stands for ultra-low-voltage 9-transistor SRAM cell. The bit cell uses less write power because to the lower activity factor and the breaking of the feedback loop between the cross-coupled inverters during a write operation [49]. It has a higher read static noise margin than a normal 6T SRAM cell in terms of minimum area (by 3.09×). At isoarea, LP9T has a higher static margin for write operations than 8T (minimum area).

6.5.5 10T SRAM Cells

The implementation of a 10-T (transistor) SRAM cell provides improved static noise margin (SNM) and lower power. Figure 6.6 shows the circuit diagram of 10T Static RAM cell with low power and improved SNM. The 10-T SRAM boosts the SNM with a single-bit line and dynamic feedback control by using less power. Furthermore, the use of sleep transistors drops power consumption [50].

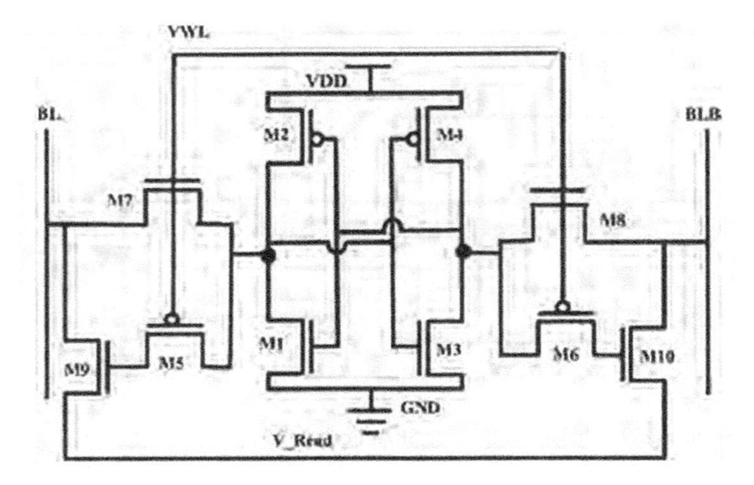

Figure 6.6 Circuit diagram for 10T SRAM Cell with improved SNM.

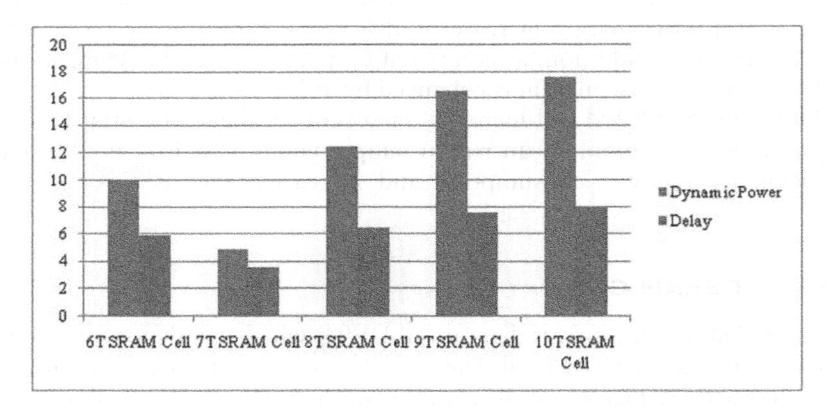

Figure 6.7 Performance comparison of various SRAM cell configurations based on dynamic power and delay.

6.6 PERFORMANCE COMPARISON OF SRAM CELL DESIGN

The behavior of GNR-based designs may be determined based on performance comparisons of GNR-based designs to CMOS, FinFET, and CNTFET-based designs [51–53]. Some of the measures used to compare efficiency include power utilization, total power dissipation, average latency, and leakage current [54, 55]. Standford and nanohub provided the open-source model files [56].

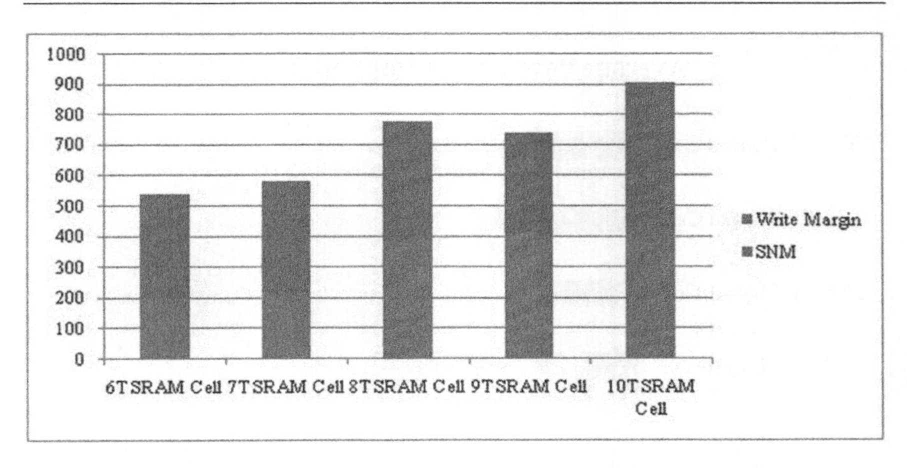

Figure 6.8 Performance comparison of Various SRAM cell configurations based on write margin and SNM.

Table 6.1 Performance Analysis of 6TSRAM Cell in Different Technologies

Parameters	*CMOS-Based Cell*	*CNTFET-Based Cell*	*FinFET-Based Cell*	*GNRFET-Based Cell*
Average Power Consumption	16.61 nW	6.21 nW	10.23 nW	8.23 nW
Total Voltage Source Power Dissipation	5.51 nW	32.9 pW	34.1 nW	0.401 nW
Average Delay	0.29 us	2.50 ns	4.65 ns	3.75 ns
Power-Delay Product (Joule)	4.8×10^{-15}	1.55×10^{-17}	4.75×10^{-17}	3.08×10^{-17}

The performance comparison of various configurations (different transistors) of SRAM cells based on dynamic power and delay is shown in Figure 6.7.

The different configurations of CMOS-based SRAM cells on write margin and SNM are shown in Figure 6.8.

The performance of GNRFET-based SRAM cells is compared to that of CMOS-, FinFET-, and CNTFET-based SRAM cells in Table 6.1 [8].

From the Figure 6.9, it is clear that a CMOS-based 6TSRAM cell consumes more power for both read and write operations at 32nm technology node. The CNTFET-based design consumes less power compared with other promising alternate devices, such as GNRFET and FinFET.

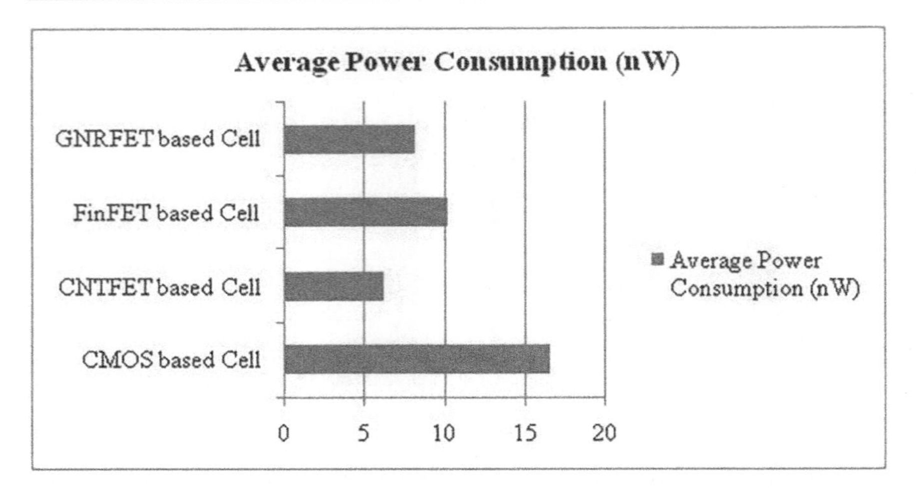

Figure 6.9 Performance Comparison of 6T SRAM cell under different technologies.

6.7 CONCLUSION

GNRFETs have dissipated less power than CMOS despite the fact that CMOS is irreplaceable. Beyond 32 nm, CNTFET dissipates less power than CMOS. Circuits based on GNRFETs are more powerful than circuits based on CMOS beyond 32 nm. The prospect of GNR in VLSI is intriguing, and it might lead the way for MOSFET technological advancements. As a result, GNRs might be used as a CMOS transistor substitute in devices larger than 32 nm.

REFERENCES

[1] Natarajamoorthy, Mathan, Jayashri Subbiah, Nurul Ezaila Alias, and Michael Loong Peng Tan, "Stability Improvement of an Efficient Graphene Nanoribbon Field-Effect Transistor-Based SRAM Design", Journal of Nanotechnology, Vol. 2020, Article ID 7608279, PP. 7, 2020. https://doi.org/10.1155/2020/7608279.

[2] International Technology Roadmap for Semiconductors (ITRS). San Jose, CA: Semiconductor Industry Association, 2007.

[3] Kumar, Chanchal, Avinash Sharan Mishra, and Vijay Kumar Sharma, "Leakage Power Reduction in CMOS Logic Circuits Using Stack ONOFIC Technique", Proceedings of the Second International Conference on Intelligent Computing and Control Systems (ICICCS 2018), PP. 1363–1368.

[4] Joshi, Shital, and Umar Alabawi, "Comparative Analysis of 6T, 7T, 8T, 9T, and 10T Realistic CNTFET Based SRAM", Journal of Nanotechnology, Vol. 2017, Article ID 4575013, PP. 9, 2017. https://doi.org/10.1155/2017/4575013.

[5] Raja, Boopathi G., and M. Madheswaran, "Design and Performance Comparison of 6-T SRAM Cell in 32nm CMOS, FinFET and CNTFET

Technologies", International Journal of Computer Applications, Vol. 70, No. 21, PP. 1–6, May 2013.

[6] https://www.eetimes.com/moores-law-dead-by-2022-expert-says/.

[7] Raja, G. Boopathi, and M. Madheswaran, "Design and Analysis of 5-T SRAM Cell in 32nm CMOS and CNTFET Technologies", International Journal of Electronics and Electrical Engineering, Vol. 1, No. 4, PP. 256–261, December 2013. https://doi.org/10.12720/ijeee.1.4.256-261.

[8] Raja, G. Boopathi, and M. Madheswaran, "Performance Comparison of GNRFET Based 6T SRAM Cell with CMOS, FinFET and CNTFET Technology", International Journal of Innovative Research in Science and Engineering, 2016. Vol. 2, No. 5, PP. 197–204.

[9] Singh, S., and B. Raj, "Analytical Modeling and Simulation analysis of T-Shaped III-V Heterojunction Vertical T-FET", Superlattices and Microstructures, Elsevier, Vol. 147, PP. 106717, November 2020.

[10] Chawla, T., M. Khosla, and B. Raj, "Optimization of Double-gate Dual Material GeOI-Vertical TFET for VLSI Circuit Design", IEEE VLSI Circuits and Systems Letter, Vol. 6, No. 2, PP. 13–25, August 2020.

[11] Kaur, M., N. Gupta, S. Kumar, B. Raj, and Arun Kumar Singh, "RF Performance Analysis of Intercalated Graphene Nanoribbon Based Global Level Interconnects", Journal of Computational Electronics, Springer, Vol. 19, PP. 1002–1013, June 2020.

[12] Wadhwa, G., and B. Raj, "An Analytical Modeling of Charge Plasma Based Tunnel Field Effect Transistor with Impacts of Gate underlap Region" Superlattices and Microstructures, Elsevier, Vol. 142, PP. 106512, June 2020.

[13] Singh, S., and B. Raj, "Modeling and Simulation Analysis of SiGe Hetrojunction Double Gate Vertical T-Shaped Tunnel FET", Superlattices and Microstructures, Elsevier, Vol. 142, PP. 106496, June 2020.

[14] Singh, S., and B Raj, "A 2-D Analytical Surface Potential and Drain Current Modeling of Double-Gate Vertical T-Shaped Tunnel FET", Journal of Computational Electronics, Springer, Vol. 19, PP. 1154–1163, April 2020.

[15] Singh, S., S. Bala, B. Raj, and Br. Raj, "Improved Sensitivity of Dielectric Modulated Junctionless Transistor for Nanoscale Biosensor Design", Sensor Letter, ASP, Vol. 18, PP. 328–333, April 2020.

[16] Kumar, V., S. Kumar, and B. Raj, "Design and Performance Analysis of ASIC for IoT Applications", Sensor Letter ASP, Vol. 18, PP. 31–38, January 2020.

[17] Young, I., and D. Nikonov, "Principles and Trends in Quantum Nano-Electronics and Nano-Magnetics for Beyond-CMOS Computing 2017", in 47th European Solid-State Device Research Conference (ESSDERC), 2017, PP. 1–5.10.1109/ESSDERC.2017.8066577. 978-1-5090-5978-2.

[18] Chen, Ying-Yu, Amit Sangai, Artem Rogachev, Morteza Gholipour, Giuseppe Iannaccone, Gianluca Fiori, and Deming Chen, "A SPICE-Compatible Model of MOS-Type Graphene Nano-Ribbon Field-Effect Transistors Enabling Gate- and Circuit-Level Delay and Power Analysis under Process Variation", in IEEE Transactions on Nanotechnology, 2015.

[19] Javey, A., J. Guo, D.B. Farmer, Q. Wang, D. Wang, R.G. Gordon, M. Lundstrom, and H. Dai, "Carbon Nanotube Field-Effect Transistors with Integrated Ohmic Contacts and High-K Gate Dielectrics", Nano Letters, Vol. 4, No. 3, PP. 447–450, 2004.

[20] Chilstedt, S., C. Dong, and D. Chen, Carbon Nanomaterials Transistors and Circuits, Transistors: Types, Materials and Applications, Nova Science Publishers, New York, NY, 2010.
[21] de Heer, W., C. Berger, E. Conrad, P. First, R. Murali, and J. MeindI, "Pionics: The Emerging Science and Technology of Graphene-Based Nanoelectronics", in Electron Devices Meeting (IEDM), IEEE International, 2007, PP. 199–202.
[22] Levendorf, M.P., C.S. Ruiz-Vargas, S. Garg, and J. Park, "Transfer-Free Batch Fabrication of Single Layer Graphene Transistors", Nano Letters, Vol. 9, No. 12, PP. 4479–4483, 2009.
[23] Wessely, P.J., F. Wessely, E. Birinci, U. Schwalke, and B. Riedinger, "Transfer-Free Fabrication of Graphene Transistors", Journal of Vacuum Science and Technology B, Vol. 30, No. 3, 2012.
[24] Betti, A., G. Fiori, and G. Iannaccone, "Strong Mobility Degradation in Ideal Graphene Nanoribbons Due to Phonon Scattering", Applied Physics Letters, Vol. 98, No. 21, PP. 212111, 2011.
[25] Son, `Y.-W., M.L. Cohen, and S.G. Louie, "Energy Gaps in Grapheme Nanoribbons", Physical Review Letters, Vol. 97, PP. 216803, 2006.
[26] Wang, X., Y. Ouyang, X. Li, H. Wang, J. Guo, and H. Dai, "Room Temperature All-Semiconducting Sub-10-nm Graphene Nanoribbon Field Effect Transistors", Physical Review Letters, Vol. 100, PP. 206803, May 2008.
[27] Raza, H., "Zigzag Graphene Nanoribbons: Bandgap and Midgap State Modulation", Journal of Physics: Condensed Matter, Vol. 23, No. 38, PP. 382203, 2011.
[28] Taba, Monica, and Gerhard Klimeck, "Investigation of the Electrical Characteristics of Triple-Gate FinFETs and Silicon-Nanowire FETs", 2006. https://nanohub.org/resources/1715.
[29] Wadhwa, G., and B. Raj, "Design and Performance Analysis of Junctionless TFET Biosensor for High Sensitivity", IEEE Nanotechnology, Vol. 18, PP. 567—574, 2019.
[30] Wadhera, T., D. Kakkar, G. Wadhwa, and B. Raj, "Recent Advances and Progress in Development of the Field Effect Transistor Biosensor: A Review", Journal of Electronic Materials, Springer, Vol. 48, No. 12, PP. 7635–7646, December 2019.
[31] Singh, S., and B. Raj, "Design and Analysis of Hetrojunction Vertical T-Shaped Tunnel Field Effect Transistor", Journal of Electronics Material, Springer, Vol. 48, No. 10, PP. 6253–6260, October 2019.
[32] Goyal, C., J.S. Ubhi, and B. Raj, "A Low Leakage CNTFET Based Inexact Full Adder for Low Power Image Processing Applications", International Journal of Circuit Theory and Applications, Wiley, Vol. 47, No. 9, PP. 1446–1458, September 2019.
[33] Sharma, S.K., B. Raj, and M. Khosla, "Enhanced Photosensivity of Highly Spectrum Selective Cylindrical Gate In1-xGaxAs Nanowire MOSFET Photodetector", Modern Physics Letter-B, Vol. 33, No. 12, PP. 1950144, 2019.
[34] Singh, J., and B. Raj, "Design and Investigation of 7T2M NVSARM with Enhanced Stability and Temperature Impact on Store/Restore Energy", IEEE Transactions on Very Large Scale Integration Systems, Vol. 27, No. 6, PP. 1322–1328, June 2019.
[35] Bhardwaj, A.K., S. Gupta, B. Raj, and Amandeep Singh, "Impact of Double Gate Geometry on the Performance of Carbon Nanotube Field Effect Transistor Structures for Low Power Digital Design", Computational and Theoretical Nanoscience, ASP, Vol. 16, PP. 1813–1820, 2019.

[36] Goyal, C., J. Subhi, and B. Raj, "Low Leakage Zero Ground Noise Nanoscale Full Adder Using Source Biasing Technique", Journal of Nanoelectronics and Optoelectronics, American Scientific Publishers, Vol. 14, PP. 360–370, March 2019.
[37] Akashe, S., S. Bhushan, and S. Sharma, "High Density and Low Leakage Current Based 5T SRAM Cell Using 45 nm Technology", International Conference on Nanoscience, Engineering and Technology (ICONSET 2011), Chennai, India, 2011, PP. 346–350. https://doi.org/10.1109/ICONSET.2011.6167978.
[38] Singh, A., M. Khosla, and B. Raj, "Design and Analysis of Dynamically Configurable Electrostatic Doped Carbon Nanotube Tunnel FET", Microelectronics Journal, Elesvier, Vol. 85, PP. 17–24, March 2019.
[39] Goyal, C., J.S. Ubhi, and B. Raj, "A Reliable Leakage Reduction Technique for Approximate Full Adder with Reduced Ground Bounce Noise", Journal of Mathematical Problems in Engineering, Hindawi, Vol. 2018, Article ID 3501041, PP. 16, 15 October 2018.
[40] Mehrabi, K., B. Ebrahimi, and A. Afzali-Kusha, "A Robust and Low Power 7T SRAM Cell Design", in 2015 18th CSI International Symposium on Computer Architecture and Digital Systems (CADS), Tehran, Iran, 2015, PP. 1–6. https://doi.org/10.1109/CADS.2015.7377782
[41] Wadhwa, G., and B. Raj, "Label Free Detection of Biomolecules Using Charge-Plasma-Based Gate Underlap Dielectric Modulated Junctionless TFET", Journal of Electronic Materials (JEMS), Springer, Vol. 47, No. 8, PP. 4683–4693, August 2018.
[42] Wadhwa, G., and B. Raj, "Parametric Variation Analysis of Charge-Plasma-Based Dielectric Modulated JLTFET for Biosensor Application", IEEE Sensor Journal, Vol. 18, No. 15, 1 August 2018.
[43] Yadav, D., S.S. Chouhan, S.K. Vishvakarma, and B. Raj, "Application Specific Microcontroller Design for IoT based WSN", Sensor Letter, ASP, Vol. 16, PP. 374–385, May 2018.
[44] Singh, G., R.K. Sarin, and B. Raj, "Fault-Tolerant Design and Analysis of Quantum-Dot Cellular Automata Based Circuits", IEEE/IET Circuits, Devices & Systems, Vol. 12, PP. 638—664, 2018.
[45] Singh, J., and B. Raj, "Modeling of Mean Barrier Height Levying Various Image Forces of Metal Insulator Metal Structure to Enhance the Performance of Conductive Filament Based Memristor Model", IEEE Nanotechnology, Vol. 17, No. 2, PP. 268–267, March 2018 (SCI).
[46] Moradi, F., M. Tohidi, B. Zeinali, and J.K. Madsen, "8T-SRAM Cell with Improved Read and Write Margins in 65 nm CMOS Technology", inL. Claesen, M.T. Sanz-Pascual, R. Reis, A. Sarmiento-Reyes, Eds.; VLSI-SoC, Internet of Things Foundations. VLSI-SoC 2014. IFIP Advances in Information and Communication Technology, Vol. 464. Springer, Cham, 2015. https://doi.org/10.1007/978-3-319-25279-7_6.
[47] Jain, A., S. Sharma, and B. Raj, "Analysis of Triple Metal Surrounding Gate (TM-SG) III-V Nanowire MOSFET for Photosensing Application", Opto-Electronics Journal, Elsevier, Vol. 26, No. 2, PP. 141–148, May 2018.
[48] Jain, N., and B. Raj, "Parasitic Capacitance and Resistance Model Development and Optimization of Raised Source/Drain SOI FinFET Structure for Analog Circuit Applications", Journal of Nanoelectronics and Optoelectronins, ASP, USA, Vol. 13, PP. 531–539, April 2018.

[49] Roy, C., and A. Islam, "Characterization of Single-Ended 9T SRAM cell", Microsystem Technologies, Vol. 26, PP. 1591–1604, 2020. https://doi.org/10.1007/s00542-019-04700-z.

[50] Grace, P.S., and N.M. Sivamangai, "Design of 10T SRAM Cell for High SNM and Low Power," in 2016 3rd International Conference on Devices, Circuits and Systems (ICDCS), Coimbatore, India, 2016, PP. 281–285. https://doi.org/10.1109/ICDCSyst.2016.7570609.

[51] Deng J., and H.S.P. Wong, "A Compact SPICE Model for Carbon Nanotube Field-Effect Transistors Including Nonidealities and Its Application-Part I: Model of the Intrinsic Channel Region", IEEE Transactions on Electron Devices, Vol. 54, No. 12, PP. 3I86–3194, December 2007.

[52] Deng J., and H.S.P. Wong, "A Compact SPICE Model for Carbon Nanotube Field-Effect Transistors Including Nonidealities and Its Application-Part II: Full Device Model and Circuit Performance Benchmarking", IEEE Transactions on Electron Devices, Vol. 54, No. 12, PP. 3195–3205, December 2007.

[53] Deng, J., and H.-S.P. Wong, "A Circuit-Compatible SPICE Model for Enhancement Mode Carbon Nanotube Field Effect Transistors," Simulation of Semiconductor Processes and Devices, 2006.

[54] Predictive technology model for 32 nm CMOS technologies. [Online]. Available at: http://www.eas.asu.edu/~ptm.

[55] Stanford University CNTFET Model Website. [Online]. Available at: http://nano.stanford.edu/model.php?id=23.

[56] nanoHUB [Online]. Availabe: http://nanohub.org.

Chapter 7

Novel Subthreshold Modeling of FinFET-Based Energy-Effective Circuit Designs

Kavita Khare, Ajay Kumar Dadoria, and Afreen Khursheed

CONTENTS

7.1 INTRODUCTION

In the past couple of years, energy dissipation has been given a considerable weightage over the chip area and switching speed parameters. The main decisive feature behind this is ever budding requirement for battery controlled high performance transportable application-specific integrated circuit (ASIC) devices. Thus, modeling such circuits, which need stringent energy restraints for a greater battery life span, is the need of hour. Relevant to this context, operating the very large-scale integration (VLSI) devices in subthreshold mode opens a doorway of opportunities for power-restrained device or circuit applications with its very low-energy consumption

DOI: 10.1201/9781003311379-7

requirements. Henceforth, the advantage of employing low down-power device operations has chiseled a forte for subthreshold circuits. Although on one hand, the subthreshold operation showcases a huge prospective toward fulfilling the very low energy requirements of transportable gadgets, but on the other hand, it invites challenges for circuit designers. These imposed challenges, resulting in a noteworthy increase in design complexity of ASICs. Ironically, very limited researchers tackle these challenges for subthreshold device design in an incorporated and inclusive way.

Moreover, with the upgrading of device technology, the dimensions of transistors are significantly scaled downward. This further pushes the ongoing demand for a high-speed and lower power device. Incorporating state-of-art fin field-effect transistor (FinFET) technology for device modeling at the subthreshold region is an impending and doable way out for highly developed and ultra-modern low down-power ASICs. Accordingly, this issue is ingeniously taken up and methodically showcased in this chapter.

7.2 SUBTHRESHOLD CIRCUIT MODELING

In the subthreshold region, ideally no current flows between the source and the drain. But in reality, because of certain non-ideal effects and parasitic a small amount of current called leakage current might flows through the device.

It is mainly observed that in the subthreshold mode of operation, the effective voltage on gate electrode $\left(V_{gatedsource}\right)$ is quite lesser or, in fact, corresponding to the threshold voltage (V_{th}) of transistor.

The current in the subthreshold mode alters in an exponential trend with $\left(V_{gatedsource}\right)$ and is expressed as

$$I_D = \frac{\mu_n C_{ox} W}{L}(\eta - 1)V_T^2 e^{\frac{V_{gs}-V_{th}}{\eta V_T}}\left(1 - e^{-V_{ds}/V_T}\right), \tag{7.1}$$

Where $\left(V_{drainsource}\right)$ is the drain to source voltage; W and L are the width and length of the transistor, respectively; C_{ox} is the oxide capacitance; and V_T is the thermal voltage. η is slope factor and is expressed as

$$\eta = 1 + \frac{C_D}{C_{ox}}, \tag{7.2}$$

where C_D is depletion-layer capacitance.

7.2.1 I_{ON} to I_{OFF} Current Ratio

I_{ON} is the subthreshold leakage charge current at $V_{GatedSource} = V_{DrainSource} = V_{DD}$ and I_{OFF} is the subthreshold leakage charge current at $V_{GatedSource} = 0$

and $V_{DrainSource} = V_{DD}$. A small I_{ON} to I_{OFF} in sub-V_{TH} circuits projects limits the utmost number of bits per line in ultra-low down-power design for memory.

7.2.2 Subthreshold Slope

Subthreshold slope (S) depends on the gate to source (V_{GS}) of the transistor. It is defined as the amount of V_{GS} required to change the subthreshold current by an order of magnitude.

$$S = 2.3V_t m, \tag{7.3}$$

where m can be defined as

$$m = \left[1 + \frac{3T_{ox}}{W_{dep}}\right]\left[1 + \frac{11T_{ox}}{W_{dep}} e^{\left[\left(\left(-\pi L_{eff}\right)/\left(2\left(W_{dep} + 3T_{ox}\right)\right)\right)\right]}\right]. \tag{7.4}$$

The value of S, which is theoretically limited to 60 mV/dec at T = 300 K, should be as small as possible to ensure the steepest subthreshold characteristics.

L_{eff} is the effective channel length, W_{dep} is the depletion width, and ε_{si} and ε_{ox} are the dielectric constants of silicon and oxide layers. T_{ox} is the oxide thickness.

Scaling of metal–oxide semiconductor (MOS) device improves the performance of the digital circuits, but as transistor size reduces, it suffers from undesirable SCE, which results in drain-induced barrier lowering (DIBL), a high electric field across the drain region, which lowers the barrier height; an increase in subthreshold slope and I_{OFF} current, which results in an increase of leakage current known as subthreshold leakage; and significant gate oxide leakage. To overcome the problem of SCEs, a novel device is investigated to reduce the leakage power: a multigate FET known as FinFET.

The International Technology Roadmap for Semiconductors (ITRS) in its recent prediction of "Beyond Moore's Law" stated that in near future, the transistor would eventually stop shrinking. This statement has been construed as death knell regarding Moore's law. This is owing to the fact that the manufacturing industry is unable to protract the scaling of CMOS. Therefore, as Si-MOSFET scaling is already reached its limiting value for a number of reasons, including very elevated leakage charge currents, elevated power density, huge parametric variations and reduced gate control, thereby making it not as appropriate for the upcoming next-generation ultra-low-power and ultra-high-speed applications. These shortcomings of the existing technology push the investigations for an assortment of alternatives to keep Moore's law "get going". Furthermore, compared to conventional silicon MOSFETs, various alternative technologies, such as FinFET [1], carbon nanotube field-effect transistor (CNT-FET) [2] and graphene-nanoribbon field-effect transistor (GNRFET) [3], were proposed. Of all these alternatives, FinFETs were found to offer improved scalability and performance with reduced SCE and superior charge carrier mobility.

However, further delving into the issue of power dissipation; it can be stated that totality power utilization in a circuit is mainly due to three key reasons, that is, dynamic-power consumption, short-circuit-power consumption and static or leakage power consumption. Two type of scaling is most popular in VLSI circuits technology constant field scaling and constant voltage scaling. Power supply holds a direct proportionality to dynamic power consumption, so reducing power significantly improves the overall performance of circuit. Scaling the supply of power in IC is for the most widely adopted scheme for reducing the dynamic and short-circuit power utilization, at the trade-off of a hike in propagation delay [4] due to lesser transistor current and by diminishing clock frequency [5]. Due to this supply, the voltage of the critical path is not altered because of speed constrain of the design. To rule out this problem, it is advised to use multivalued-V_{DD} system; in this scenario, the critical path is dispensed with regular power supply (V_{DDH}), and the noncritical trail is dispensed by the downscale of supply voltage (V_{DDL}) [6]. This creates diverse voltage levels [7], and to facilitate the communication among each other, a voltage-level converter, termed a level shifter (LS), is employed as an interface between the circuit in a low-power circuit design.

7.3 EMERGING NANOMETER SUBTHRESHOLD DEVICE TECHNOLOGIES

7.3.1 Single-Electron Transistors

Single-electron transistors, abbreviated SETs, shown in Figure 7.1 are basically nano-devices consisting of small islands. These islands are conducting quantum dots, connected to a source, and drained with tunnels. In addition to this, they are capacitive when coupled to the gate junction. SET devices are based on the quantum phenomenon of the tunneling effect. SETs are named on the basis of their mode of operation. Each time an electron is appended in it, the transistor switches ON/OFF. SETs are not widely accepted due to their implementation restrictions, such as precise room temperature, arbitrariness in background charge, the phenomenon of co-tunneling, the nonexistence of appropriate lithography, and being unable to link properly with the outside circuit, although very much research is still going on in this area for overcoming the said restrictions. Hence, other options are discussed in the following sections.

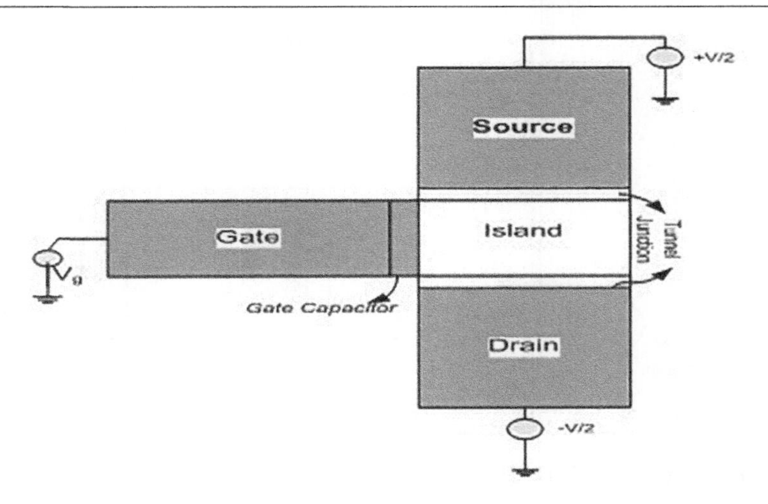

Figure 7.1 Structure of an SET.

7.3.2 CNTFET

CNTFETs consist of single-walled carbon nanotube as the conducting channel for the FET [8]. The mode of operation of CNTFET is very much similar to a conventional MOSFET [9]. In comparison to MOSFETs, the novel CNTFETs show superiority by exhibiting excellent control over channel formation, good threshold voltage controllability [10] high current density and high transconductance. Due to high electron mobility, CNTFETs offer greater current velocity than MOSFETs. Figure 7.2. shows the CNTFET structure [11] with a carbon nanotube forming the channel region; channel region is undoped, but other regions are heavily doped. The heavily doped regions are termed the source and drain regions.

Some of the limitations of CNTFETs include difficulty in controlling the chirality of the nanotubes, the degradation in CNTs on exposure to oxygen, and the influence of high temperature, and under high electric fields, CNTs become highly unreliable. Although research is going on for the thermal stability of CNTFETs, yet another drawback with CNTFETs that restricts their use in the industry is that ASICs have an excessively high production cost and cumbersome mass production.

7.3.3 GNRFETs

GNRFETs are also carbon-based FETs that are very much similar to CNTFETs. GNRFETs have better performance, lower sensitivity regarding the inconsistency of channel chirality, and reduced leakage problems because of interband tunneling. GNRFETs are classified as GNR-MOSFETs and

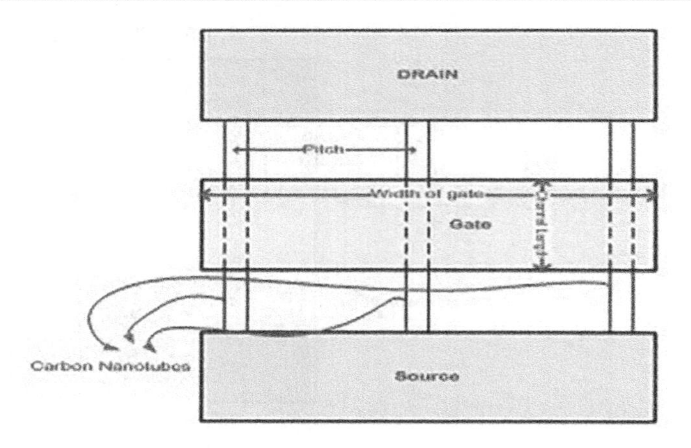

Figure 7.2 Structure of a CNTFET.

GNR-schottky barrier field effect transistors. GNR-MOSFETs as compared to SB-MOSFETs has a greater achievable maximum I_{ON}/I_{OFF} ratio, larger on-current, high transconductance, and better saturation behavior with less output conductance. Compared to traditional MOSFETs, GNRFETs have better switching and high-frequency performance. The drawback with GNRFETs that restrict its use in the industry is that in the ASIC, the edge roughness of GNR can result in a greater OFF current and a smaller ON current. Moreover, a negative impurity disturbs the carrier transportation of GNRFETs due to the locally increased electrostatic potential.

7.3.4 FinFETs

As mentioned, the scaling of CMOS devices improves the performance of the digital circuits, but today, the scaling of MOSFETs has slowed. As technology scales down, transistor sizes are reduced but suffer from undesirable SCE, which results in DIBL; a high electric field across the drain region, which lowers the barrier height; an increase in the subthreshold slope and I_{OFF} current, which results in an increase of leakage current known as subthreshold leakage; and significant gate oxide leakage. To overcome from the problem of SCEs, a novel device is investigated to reduce the leakage power [12–14].

FinFETs come under the category of a multigate transistor. It was introduced by researchers at the University of California, Berkeley to define a nonplanar, double-gate transistor with a conducting channel wrapped by a thin silicon FIN. This forms the body of the device. FinFETs are widely used device technology these days due to faster logic and reduced dynamic power consumption and static leakage current. Vertical channel structures (gates) that resemble fish fins are used in FinFETs. These fins are typically lightly doped or undoped; therefore, the improvement in carrier mobility and reduction in the doping fluctuation in a double gate (DG) is compared to the bulk transistor. With this new device technology, the needs of nanometer-era applications can be accomplished, and the performance of circuits can be enhanced. This manuscript mainly focuses on Fin-FET modeling due to its advantages and its promising electrical characteristics to serve as a replacement for conventional CMOS technology. The following section discusses the FinFET structure and device modeling in detail [15–17].

7.4 FINFETS: STATE-OF-THE-ART TECHNOLOGY

As mentioned, the chief benefit of the Fin-FET arrangement is the fabrication of a dual gate by means of a solo lithography and etching stage. In FinFETs, the gate is easily wounded over the silicon fin on account of the fact that the front gate and the back gate have different doping profiles, so both the gate can be made to operate autonomously as per the requirement. One of the foremost challenges while making of FinFET over bulk CMOS, it has elevated the current drive by plummeting parasitic resistance. It's required to reengineered source and drain regions [18–20].

FinFET has a three-dimensional framework as revealed in Figure 7.3. In Fin-FETs, the conductive channel is created at a 90-degree angle to the wafer plane, and the charge current flows parallel to the wafer plane; henceforth, the device is termed quasi-planar. FinFET devices have excellent performance improvement and low-power characteristics that result in a higher I_{ON}/I_{OFF} ratio, suppress the leakage current, and increase the switching speed of the circuit. Due to the vertical gate structure of FinFETs, the width is quantized, and fin height is determined by minimum transistor width (W_{min}) as shown in Figures 7.3a and 7.3b. When the two gates of a single FinFET are tied together, W_{min} is the channel's effective width (W_{min}), and the channel's effective length (L_{eff}) is determined by the given equations

$$W_{min} = 2H_{fin} + T_{fin}, \tag{7.5}$$

$$L_{eff} = L_{gate} + 2 \times L_{ext}, \tag{7.6}$$

where H_{fin} is the stature height of the fin, T_{fin} is the thickness of the silicon body, and L_{ext} is the lean-to of the fin from the gate to the source or the drain terminal. In order to reduce the effect of the shorter channel and enrich the area efficiency in FinFET, fin thickness is to a great extent smaller than fin height. Table 7.1 shows the parameters to be considered during simulation N-FinFET and P-FinFET.

FinFETs have better optimization of the subthreshold leakage current and improve the performance to mitigate the power consumption. Depending on the arrangement of the gate over the fin, which is controlled independently

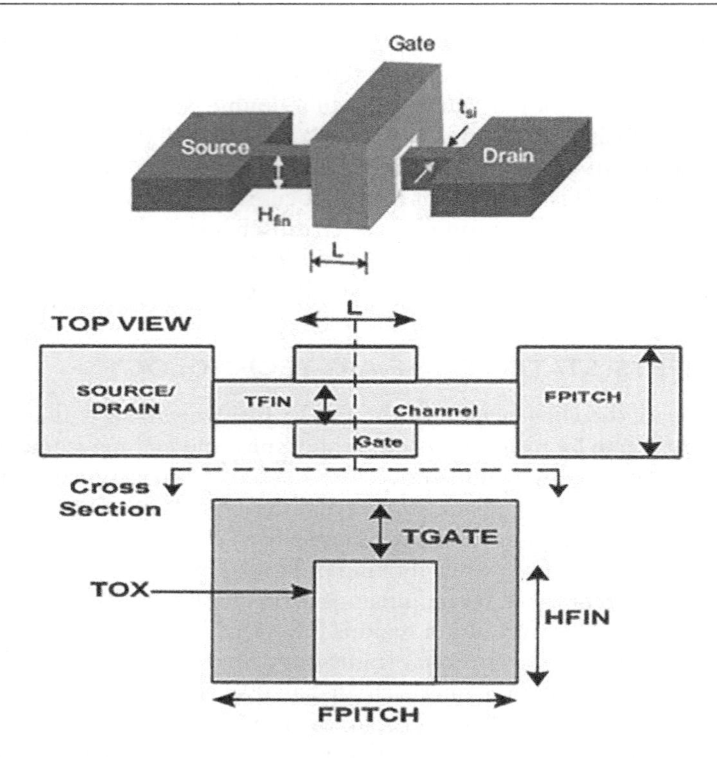

Figure 7.3 (a) FinFET three-dimensional view of single fin; (b) cross section and top view of FinFET.

with different supply voltage, as shown in Figure 7.4, FinFETs work in three different modes according to the supply of the front and back gates, namely, short-gate mode, low-power mode, and independent gate mode [21–23].

FinFET can be categorized in two symmetric double gate, which has three terminal sources, drain and gate as shown in Figure. 7.4a when both front and back gates are tied together ($V_{FG} = V_{BG} = V_{dd}$), we apply same voltage to both the gate is known as SG mode or symmetric double gate (3T) as shown in Figure 7.4b. In independent gate mode or asymmetric double-gate front gate and back gate are at different potentials ($V_{FG} \neq V_{BG}$); both gates are asymmetric as shown in Figure 7.4b, and both gates are connected independently with various configuration. Asymmetric dual gates have excellent control over the channel, which reduces I_{OFF}, current, increases I_{ON} current, achieve ideal subthreshold slop, and so on. Asymmetric DGMOS with independent biasing of the front gate and the back gate, which helps lower the dynamic threshold voltage, which helps lower the I_{OFF} and increase the I_{ON}

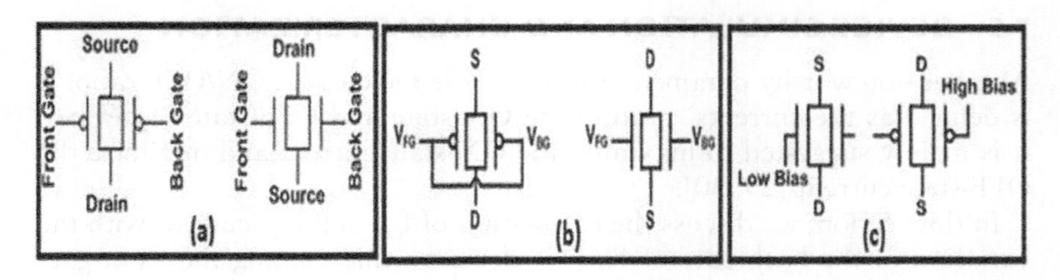

Figure 7.4 (a) DGPMOS and DGNMOS FinFET structure (b) SG mode ((V_{FG}=V_{BG})), IG mode (V_{FG}≠V_{BG}) configuration of Fin-FET, (c) LP Mode FinFET.

Table 7.1 Parameters of H℘ and LSTP Modes of a FinFET Device

Technology Parameters		*Technology*				
		7nm	*10nm*	*14nm*	*16nm*	*20nm*
H_{FIN} (nano, m)		18	21	23	26	28
W_{FIN} (nano, m)		11	14	18	20	24
L (nano, m)		7	10	14	16	20
V_{DD} (V)		0.7	0.75	0.8	0.85	0.9
Work Function HP (eV)	NFET Type	4.419	4.420	4.420	4.41	4.38
	PFET Type	4.732	4.736	4.75	4.76	4.80
Work Function LSTP (eV)	NFET Type	4.596	4.600	4.60	4.58	4.56
	PFET Type	4.559	4.559	4.57	4.59	4.62

current with low gate capacitance, to achieve high design flexibility at the circuit level [24].

7.4.1 High-Performance Model

In the high-performance (H℘) model, the threshold voltage (V_{TH}) of the transistor is lower than normal value to improve the performance of the circuit [25, 26].

7.4.2 LSTP Model

In LSTP model, the threshold voltage (V_{TH}) of the transistor is higher than normal to mitigate the leakage power, but there is degradation in the performance of the circuit.

Table 7.1 mentions the diverse parameters needed to put up the model of FinFET from 7nm to 20nm technology with variation of the power supply voltage with technology scaling [27, 28].

7.5 DEVICE SIMULATION AND CHARACTERIZATION

Another noteworthy parameter of FinFETs is the current ON/OFF ratio. It is defined as the currents' ratio in the ON-state and OFF-state of devices. It is mainly suggested to maximize the ON-state current and minimize the OFF-state current [29, 30].

In this section, we discuss the calculation of I_{ON} and I_{OFF} current with the variation of the back gate of FinFET devices while making the front-gate voltage constant. All the simulation is performed at 20nm, 16nm, 14nm, 10nm, and 7nm by employing FinFET technology in H℘ and LSTP model for optimizing the power expenditure by using an HSPICE simulator, with deviations on all the parameters with the technology scaling [31–34].

In Table 7.2, we calculated the I_{ON} and I_{OFF} current of an N-type FinFET and a P-type FinFET from 20nm to 7nm technology. We observe from the table that as we apply back-gated bias from −0.2V to −0.4V, there is no variation in I_{ON} and I_{OFF} current for the same technology, but as we proceed from 20nm to 7nm, the I_{ON} current depreciates from 1.702 μA to 1.569 μA. This depreciation is minute with the scaling technology, but the reduction of the I_{OFF} is greater, from 23.96 nA to 36.28 nA. If we reduce the I_{OFF} current, then the driving capability of the transistor also increases. Table 7.3 also follows the same trend; there is a reduction in the I_{ON} current from 1.418μA to 1.591μA. This reduction is quite small with the scaling of technology, but the reduction of the I_{OFF} is greater, from 21.17nA to 32.65nA, if we reduce the I_{OFF} current in the H℘ model of FinFET technology. In Table 7.4, there is an increase in the I_{ON} current from 1.702 μA to 526.6 μA, which increases the driving capability of the transistor, but the I_{OFF} current is also reduced from microampere to picoampere—23.87 pA to 34.67 pA—but as we move from 20nm to 7nm, there is a smaller reduction of the I_{OFF} current in the LSTP model for N-FinFET. In Table 7.4, the same trend for P-FinFET is followed: The I_{ON} current increases from 455.7 μA to 725.0 μA in the LSTP model. From Tables 7.2 through 7.5, when the back gate is biased, there is no variation in the I_{ON} and I_{OFF} currents in an N-FinFET and a P-FinFET in both the H℘ and LSTP models of FinFET technology [35–37].

Table 7.2 Computation of I_{ON} and I_{OFF} Current in an N-FinFET by Means of the H℘ Model

VFG	*VFG = VBG*		*VBG = 0*		*VBG = −0.2*		*VBG = −0.4*	
	I_{ON} (μA)	I_{OFF} (nA)	I_{ON} (μA)	I_{OFF} (nA)	I_{ON} (μA)	I_{OFF} (nA)	I_{ON} (μA)	I_{OFF} (nA)
7nm	1.569	23.96	1.569	23.96	1.569	23.96	1.569	23.96
10nm	1.631	27.58	1.631	27.58	1.631	27.58	1.631	27.58
14nm	1.601	30.06	1.601	30.06	1.601	30.06	1.601	30.06
16nm	1.601	33.90	1.601	33.90	1.601	33.90	1.601	33.90
20nm	1.702	36.28	1.702	36.28	1.702	36.28	1.702	36.28

Table 7.3 Computation of I_{ON} and I_{OFF} current in a P-FinFET by Means of the H℘ Model

VFG	VFG = VBG		VBG = 0		VBG = −0.2		VBG = −0.4	
	I_{ON} (μA)	I_{OFF} (nA)	I_{ON} (μA)	I_{OFF} (nA)	I_{ON} (μA)	I_{OFF} (nA)	I_{ON} (μA)	I_{OFF} (nA)
7nm	1.418	21.17	1.418	21.17	1.418	21.17	1.418	21.17
10nm	1.508	24.72	1.508	24.72	1.508	24.72	1.508	24.72
14nm	1.448	26.65	1.448	26.65	1.448	26.65	1.448	26.65
16nm	1.486	30.01	1.486	30.01	1.486	30.01	1.486	30.01
20nm	1.591	32.65	1.591	32.65	1.591	32.65	1.591	32.65

Table 7.4 Computation of I_{ON} and I_{OFF} current in an N-FinFET by Means of the LSTP Model

VFG	VFG = VBG		VBG = 0		VBG = −0.2		VBG = −0.4	
	I_{ON} (μA)	I_{OFF} (pA)	I_{ON} (μA)	I_{OFF} (pA)	I_{ON} (pA)	I_{OFF} (pA)	I_{ON} (μA)	I_{OFF} (pA)
7nm	526.6	23.87	526.6	23.87	526.6	23.87	526.6	23.87
10nm	653.5	27.51	653.5	27.51	653.5	27.51	653.5	27.51
14nm	716.4	29.21	716.4	29.21	716.4	29.21	716.4	29.21
16nm	761.4	33.53	761.4	33.53	761.4	33.53	761.4	33.53
20nm	814.3	34.67	814.3	34.67	814.3	34.67	814.3	34.67

Table 7.5 Computation of I_{ON} and I_{OFF} current in a P-FinFET by Means of the LSTP Model

VFG	VFG = VBG		VBG = 0		VBG = −0.2		VBG = −0.4	
	I_{ON} (μA)	I_{OFF} (pA)	I_{ON} (μA)	I_{OFF} (pA)	I_{ON} (pA)	I_{OFF} (pA)	I_{ON} (μA)	I_{OFF} (pA)
7nm	455.7	21.78	455.7	21.78	455.7	21.78	455.7	21.78
10nm	553.7	24.84	553.7	24.84	553.7	24.84	553.7	24.84
14nm	622.1	27.34	622.1	27.34	622.1	27.34	622.1	27.34
16nm	679.5	31.88	679.5	31.88	679.5	31.88	679.5	31.88
20nm	725.0	32.47	725.0	32.47	725.0	32.47	725.0	32.47

Scaling of the transistor is a prime thrust for development in semiconductor industries. In this work, we have evaluated the impact of I_{ON} and I_{OFF} current. This work is done on the H℘ and LSTP models for FinFETs for 20nm, 16nm, 14nm, 10nm, and 7nm predictive technology model–multigate (PTM-MG) as shown in Table 7.6. The PTM-MG model is developed by Berkeley Short-channel IGFET Model-Compact Model for Multi-Gate Transistors for scaling multigated devices. Here, the impact is on the I_{ON} and I_{OFF} current with a variation of fins.

Table 7.6 Results of I_{ON} and I_{OFF} for Multifins Using the H℘ and LSTP Model for 20nm to 7nm

Modes	*Tech*	I_{ds}	*Fin = 1*	*Fin = 4*	*Fin = 8*	*Fin = 16*	*Fin = 32*
H℘ NFET	20nm	I_{ON} (μA)	88.41	353.6	707.3	1414	2829
		I_{OFF} (nA)	6.555	26.22	52.44	104.8	209.7
H℘ PFET		I_{ON} (μA)	78.29	313.1	626.3	1252	2505
		I_{OFF} (nA)	6.338	25.35	50.17	101.4	202.8
LSTP NFET	20nm	I_{ON} (μA)	51.48	205.9	411.8	823.6	1647
		I_{OFF} (nA)	6.262	25.04	50.09	100.1	200.3
LSTP PFET		I_{ON} (μA)	45.91	183.6	326.2	734.5	1469
		I_{OFF} (pA)	6.200	24.88	49.75	99.51	199.0
H℘ NFET	16nm	I_{ON} (μA)	91.05	364.2	728.4	1456	2913
		I_{OFF} (nA)	5.949	23.79	47.59	95.18	190.3
H℘ PFET		I_{ON} (μA)	80.30	321.2	642.4	1284	2569
		I_{OFF} (nA)	5.693	22.77	45.54	91.09	182.2
LSTP NFET	16nm	I_{ON} (μA)	51.27	205.0	410.1	820.3	1640
		I_{OFF} (pA)	5.813	23.25	46.50	93.00	186.0
LSTP PFET		I_{ON} (μA)	45.70	182.8	365.6	731.2	1462
		I_{OFF} (nA)	5.924	23.69	47.38	94.77	189.5
H℘ NFET	14nm	I_{ON} (μA)	94.14	376.5	753.1	1506	3012
		I_{OFF} (nA)	5.204	20.81	41.63	83.27	166.5
H℘ PFET		I_{ON} (μA)	86.49	345.9	691.9	1383	2767
		I_{OFF} (nA)	5.014	20.05	40.11	80.22	160.4
LSTP NFET	14nm	I_{ON} (μA)	49.78	199.1	398.2	796.5	1593
		I_{OFF} (pA)	4.919	19.67	39.35	78.70	157.4
LSTP PFET		I_{ON} (μA)	45.65	182.6	365.2	730.4	1460
		I_{OFF} (pA)	4.970	19.87	39.75	79.51	159.0
H℘ NFET	10nm	I_{ON} (μA)	89.97	359.1	719.8	1439	2879
		I_{OFF} (nA)	4.704	18.17	37.63	75.27	150.5
H℘ PFET		I_{ON} (μA)	82.66	330.6	661.3	1322	2645
		I_{OFF} (nA)	4.584	18.33	36.67	73.34	146.6
LSTP NFET	10nm	I_{ON} (μA)	43.78	175.1	350.1	700.6	1401
		I_{OFF} (pA)	4.475	17.89	35.79	71.59	143.1
LSTP PFET		I_{ON} (μA)	39.27	157.1	314.1	628.3	1256
		I_{OFF} (pA)	4.448	17.79	35.57	71.15	142.3
H℘ NFET	7nm	I_{ON} (μA)	81.94	327.9	655.9	1311	2623
		I_{OFF} (nA)	3.973	15.87	31.75	63.50	127.0
H℘ PFET		I_{ON} (μA)	72.20	288.8	577.6	1155	2310
		I_{OFF} (nA)	3.824	15.29	30.58	61.16	122.3

Modes	Tech	I_{ds}	Fin = 1	Fin = 4	Fin = 8	Fin = 16	Fin = 32
LSTP NFET	7nm	I_{ON} (μA)	35.16	140.6	281.3	562.6	1125
		I_{OFF} (pA)	3.782	15.12	30.25	60.50	121.0
LSTP PFET		I_{ON} (μA)	31.47	125.9	251.7	503.5	1007
		I_{OFF} (pA)	3.762	15.04	30.09	60.19	120.3

In Tables 7.7 and 7.8, the dynamic power and the leakage power are measured on the basic gates from 20nm to 7nm technology. Tables 7.9 and 7.10 show the leakage power reduction technique being introduced for a fair comparison of results. The drain-gating technique has huge potential for power mitigation in terms of dynamics or leakage [38–40].

Table 7.7 Average Power Measurement on Basic Gates

Circuits	*Dynamic Power Dissipation (nW)*				
	20nm	*16nm*	*14nm*	*10nm*	*7nm*
NOT Gate	144.0	130.6	112.4	112.5	104.7
AND Gate	70.17	63.63	55.31	73.71	96.64
NAND Gate	61.42	57.77	50.58	53.77	59.65
NOR Gate	48.57	43.80	37.70	43.35	49.22
EXOR Gate	96.88	80.47	73.13	75.16	86.82

Table 7.8 Leakage Power Measurement on Basic Gates

Circuits	*Input Vector*	*Leakage Power Dissipation (nW)*				
		20nm	*16nm*	*14nm*	*10nm*	*7nm*
NOT Gate	0	0.0109	0.124	0.462	4.939	12.29
	1	0.0041	0.023	0.099	0.227	5.941
AND Gate	00	0.0022	0.092	0.339	4.292	22.89
	01	0.0050	0.213	0.786	8.744	38.46
	10	0.0045	0.179	0.649	7.641	32.52
	11	0.0037	0.171	0.663	8.446	54.05
NAND Gate	00	0.0012	0.0038	0.015	0.524	3.532
	01	0.0029	0.124	0.459	4.806	11.89
	10	0.0024	0.0899	0.322	3.737	9.854
	11	0.0036	0.0467	0.198	3.366	11.46

Continued

Table 7.8 Continued

Circuits	*Input Vector*	*Leakage Power Dissipation (nW)*				
		20nm	*16nm*	*14nm*	*10nm*	*7nm*
NOR Gate	00	0.0058	0.2485	0.913	9.136	20.61
	01	0.0041	0.0233	0.098	1.660	5.644
	10	0.0026	0.0119	0.048	0.970	3.852
	11	0.0009	0.0037	0.018	0.100	1.101
EXOR Gate	00	0.0068	0.0353	0.852	2.602	9.570
	01	0.0053	0.2138	1.184	8.054	37.09
	10	0.0053	0.2138	1.184	8.054	37.09
	11	0.0068	0.0353	0.852	2.602	9.570

Table 7.9 Average Power Measurement of Logic Gates by Drain Gating

Circuits	*Dynamic Power Dissipation (nW)*				
	20nm	*16nm*	*14nm*	*10nm*	*7nm*
NOT Gate	61.45	51.22	86.92	33.58	42.70
AND Gate	33.96	29.27	54.67	52.98	78.30
NAND Gate	73.10	28.10	32.24	40.21	42.91
NOR Gate	50.08	39.46	32.25	31.92	46.41
EXOR Gate	55.13	51.34	39.88	53.25	68.03

Table 7.10 Leakage Power Measurement of Logic Gates by on Drain Gating

Circuits	*Input Vector*	*Leakage Power Dissipation (pW)*				
		20nm	*16nm*	*14nm*	*10nm*	*7nm*
NOT Gate	0	0.35249	1.243	5.523	204.73	1735
	1	0.00911	35.66	1.714	84.97	953.5
AND Gate	00	0.07179	2.5369	11.178	392.08	3162
	01	0.07049	2.4864	11.039	406.98	3417
	10	0.11928	4.6460	20.079	615.17	4014
	11	0.05108	1.8286	8.2572	330.58	3070
NAND Gate	00	0.02349	0.8000	3.5148	125.92	1242
	01	0.03524	1.2435	5.5226	204.31	1722
	10	0.03524	1.2435	5.5230	204.50	1726
	11	0.01583	0.5847	2.7329	125.75	1335

Continued

Table 7.10 Continued

Circuits	*Input Vector*	*Leakage Power Dissipation (pW)*				
		20nm	*16nm*	*14nm*	*10nm*	*7nm*
NOR Gate	00	0.05158	1.8327	8.0598	286.97	2247
	01	0.00911	0.7782	1.7144	84.751	942.72
	10	0.00911	0.7783	1.7145	84.805	942.58
	11	0.00446	0.4257	0.7172	32.419	464.18
EXOR Gate	00	0.01583	0.5847	2.7327	146.18	1326
	01	0.05159	1.8326	8.0593	340.22	2239
	10	0.05159	1.8326	8.0593	340.22	2239
	11	0.01583	0.5847	2.7327	146.18	1326

Figure. 7.5 shows the PDP for an NFET and a PFET with back-gate biasing. The PFET PDP is higher than the NFET PDP.

As technology scales down, F and its combination circuit as shown in Figures 7.6a through 7.6d shows a greater percentage saving in dynamic power. Here drain-gating N-type and P-type transistor (DGNPT), header DGNPT, footer DGNPT, and header–footer DGNPT compared with conventional drain gating (Figure 7.6e) in the NAND gate at 20nm, 16nm, 14nm, 10nm, and 7nm in LSTP model of FinFET technology [41–43].

Figures 7.7a through 7.7c show the experimental results. The analysis of these results leads to the conclusion that as we increase the number of fins 1, 4, 8, 16, and 32, the leakage power also increases because I_{OFF} of the transistor also increases. Figure 7.7a demonstrates that there is a rapid exponentially increase in leakage current as we increase the fins at 20nm, 16nm, 14nm, 10nm, and 7nm in LSTP model of FinFET technology in conventional gates. Figures 7.7a through 7.7c show that 7nm has a larger leakage current than the 20nm, 16nm, 14nm, 10nm, and 7nm technology. Figure 7.7c shows that the DGNPT technique has the least leakage current (fW) than the conventional NAND gate (nW) and the NAND gating using drain gating (pW). The DGNPT technique shows the lowest leakage power, but the power dissipation increases exponentially as the number of fins increases because there is an increase in the OFF-state current.

The DGNPT technique and its variation are tested on a conventional NAND gate with the variation of technology shown in Figure 7.8. The three variations of DGNPT are header DGNPT, footer DGNPT, and header–footer DGNPT.

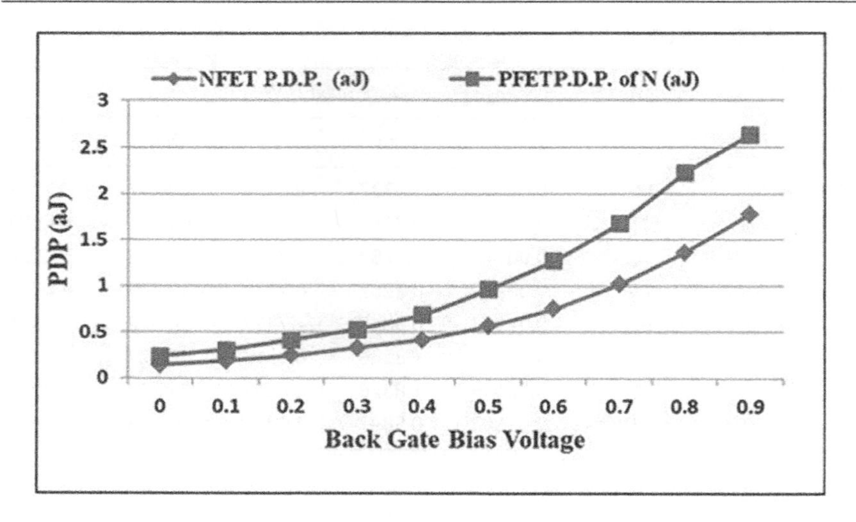

Figure 7.5 PDP of back-gate bias.

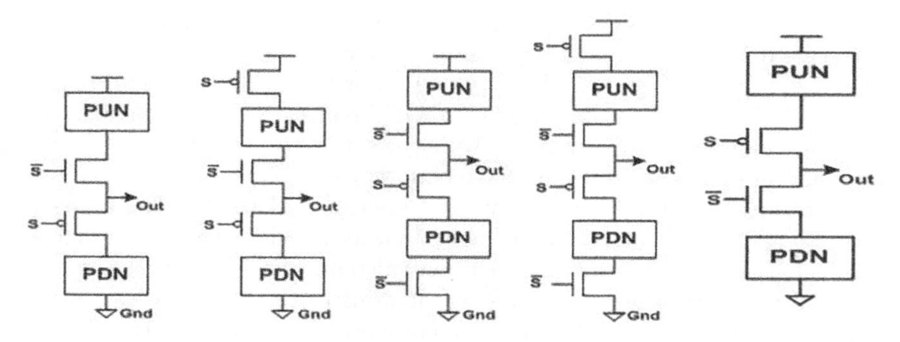

Figure 7.6 (a) DGNPT; (b) header DGNPT; (c) footer DGNPT; (d) header–footer DGNPT; and (e) conventional drain gating.

In Figure 7.9, comparisons of NAND gate DGNPT, header DGNPT, footer DGNPT, and header–footer DGNPT with conventional drain gating in NAND gating is done. From the simulation results, it is observed that drain gating technique has larger power dissipation than the DGNPT, header DGNPT, footer DGNPT, and header–footer DGNPT, and as we increase the number of fins, an exponential increase in the leakage current occurs. The DGNPT has lower leakage with respect to the drain-gating technique but has a larger leakage than the header DGNPT, footer DGNPT, and header–footer DGNPT. Minimum leakage is measured in header–footer DGNPT technique when compared with conventional drain gating, DGNPT, header DGNPT, and footer DGNPT [44].

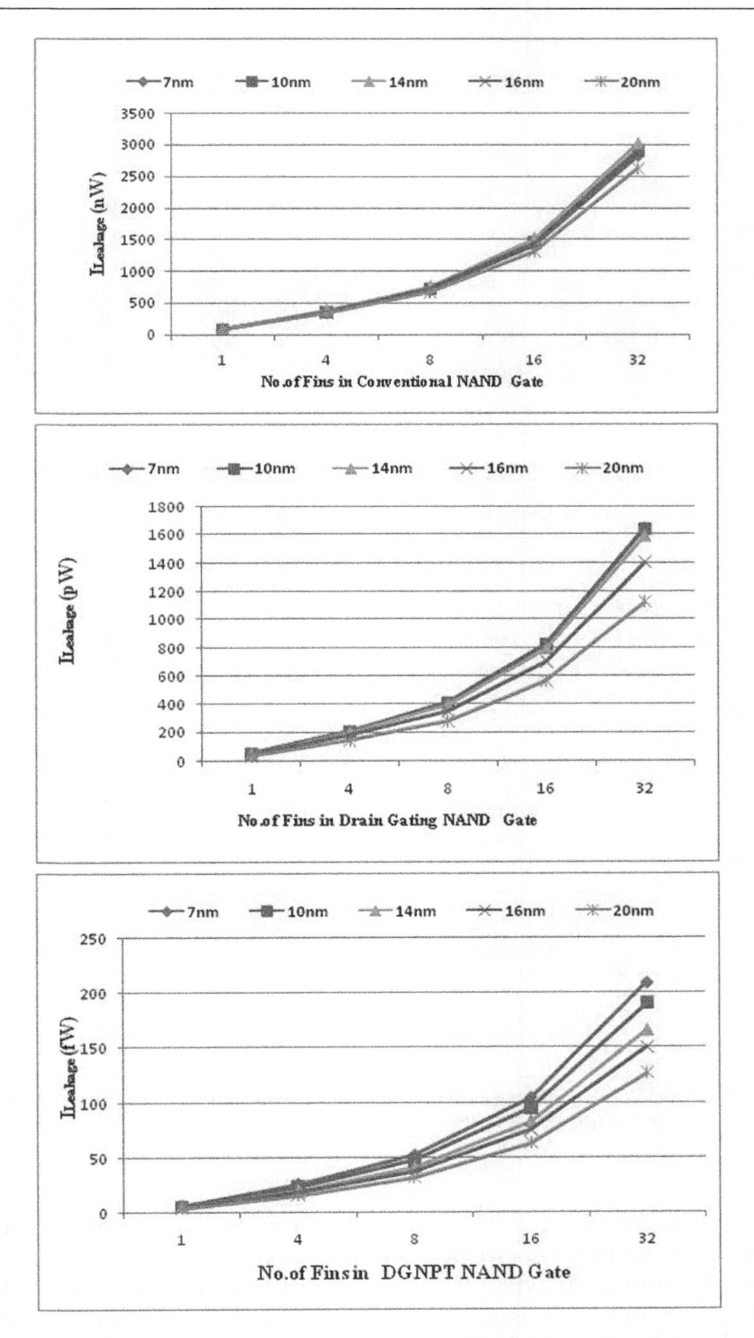

Figure 7.7 Leakage power measurement with a variation of fins: (a) Conventional NAND gate, (b) drain-gating NAND gate, and (c) DGNPT NAND gate.

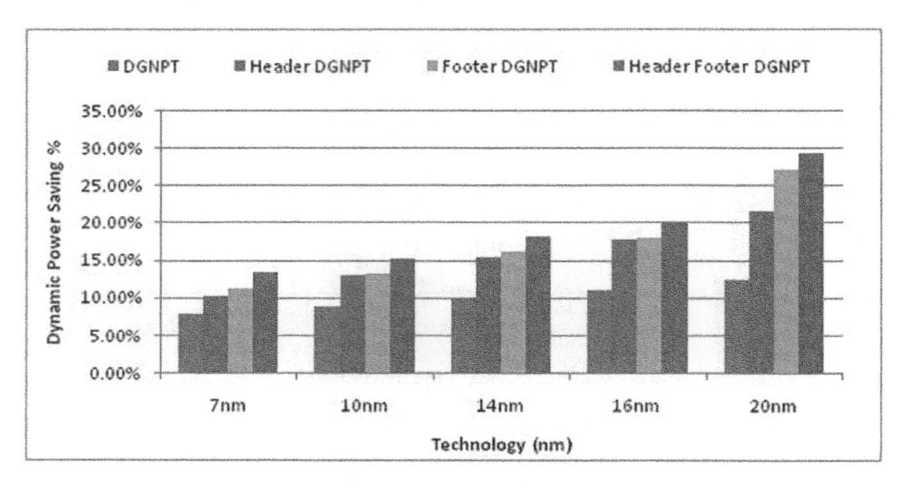

Figure 7.8 Dynamic power saving by different techniques when compared with drain gating.

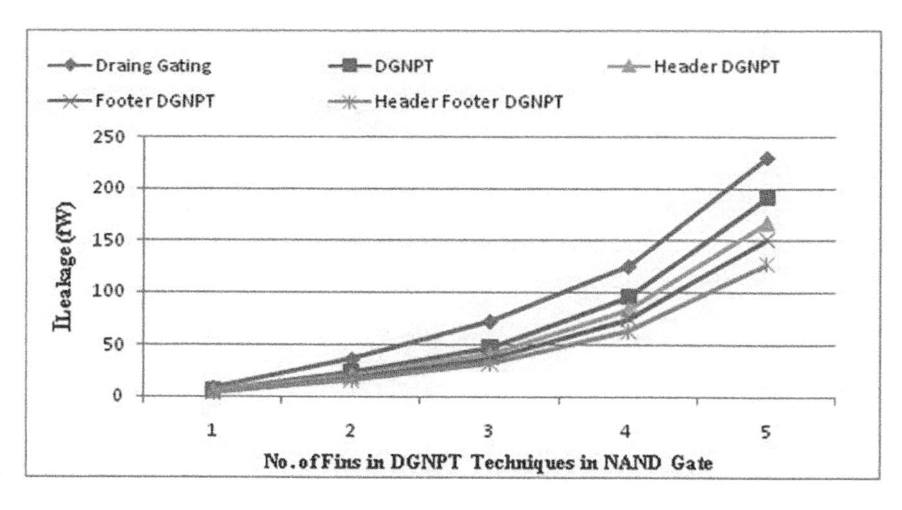

Figure 7.9 Leakage power measurement in different combinations of circuits with a variation of fins for NAND gate.

While understanding the properties of the DGNPT circuit, various designs with the mixture circuit, such as header DGNPT, footer DGNPT, and header–footer DGNPT, are implemented. Circuits along the critical path will first use only original circuit types. Then gates along critical paths are analyzed to measure their criticality. Gates with low criticality will be considered for replacement with drain-gating-based gates as long as the critical path delays are within an acceptable range. Through a number of design iterations, a highly optimized circuit type can then be determined for each gate. From the simulation results, it shows that there is a reduction in power dissipation with no increase in the critical path delay of the circuit.

7.6 CONCLUSION

The rapid advancements of technology have led to much scientific research with gigantic prospects, and from this chapter, it can be interpreted that FinFETs seem to be a superior alternative compared to conventional CMOSs and other emerging technologies due to their high performance and smaller dimensions. In this chapter, we have evaluated the electrical characteristic of FinFET circuit by calculating the I_{ON} and I_{OFF} current of N-type FinFETs and P-type FinFETs in both the H℘ and LSTP models of FinFETs from 7–20nm technology. It is observed that if we apply back-gate bias from 7–20nm technology there is no impact on I_{ON} and I_{OFF} current in both the H℘ and LSTP models. In the LSTP model, the I_{ON} is larger than in the H℘ model because it is a lower standby power library that suppresses the unwanted leakage current. In the future, FinFET-based leakage-reduction methods are preferred for low-power-consumption applications in the near-term scenario.

7.7 SUMMARY

As scaling of the transistor is a prime thrust for development in semiconductor industries, this chapter presents a simulation of the H℘ and LSTP models for FinFET versions of 20nm, 16nm, 14nm, 10nm, and 7nm PTM-MG, and the impact of I_{ON} and I_{OFF} current is evaluated with the variation of fins. The simulation shows PDP with back-gate-biasing PFET is higher than the same type of NFET.

REFERENCES

[1] Dadoria A.k., K. Khare, T.K. Gupta, and R.P. Singh, "Ultra-Low Power FinFET-Based Domino Circuits", International Journal Of Electronics, Vol. 104, PP. 952–967, 2017.

[2] Khursheed, Afreen, Kavita Khare, and Fozia Z. Haque, "Performance Tuning of VLSI Interconnects Integrated with Ultra Low Power High Speed DSM Repeaters", Journal of Nano and Optoelectronics, American Scientific Publishers, Vol. 13, No. 12, PP. 1797–1806 (10), December 2018. https://doi.org/10.1166/jno.2018.2507.

[3] Banerjee, Kaustav, Yasin Khatami, Chaitanya Kshirsagar, and S. Hadi Rasouli, "Graphene Based Transistors: Physics, Status and Future Perspectives", in Proceedings of the 2009 International Symposium on Physical design, PP. 65–66, 2009.

[4] Wong, H.-S.P., D.J. Frank, P.M. Solomon, C.H.-J. Wann, and J.J. Welser, "Nanoscale CMOS", Proceedings of IEEE, Vol. 87, PP. 537–570, 1999.
[5] Nowak, E., et al., "Turning Silicon on Its Edge", IEEE Circuits & Device Magazine, PP. 20–31, 2004.
[6] Tawfika, S.A., and V. Kursun, "FinFET Domino Logic with Independent Gate Keepers", Micro Electronics Journal, Vol. 40, PP. 1531–1540, 2009.
[7] Rasouli, S.H., H.F. Dadgour, K. Endo, H. Koike, and K. Banerjee, "Design Optimization of FinFET Domino Logic Considering the Width Quantization Property", IEEE Transactions On Electro Devices, Vol. 57, No. 11, PP. 2934–2943, 2010.
[8] Khursheed, A., K. Khare, &F.Z. Haque, "Designing High-Performance Thermally Stable Repeaters for Nano-Interconnects", Journal of Computational Electronics, Vol. 18, PP. 53–64, 2019. https://doi.org/10.1007/s10825-018-1271-0.
[9] Khursheed, A., K. Khare, and F.Z. Haque, "Designing of Ultra Low Power High Speed Repeaters for Performance Optimization of VLSI Interconnects at 32 nm", International Journal of Numerical Modelling: Electronic Networks, Devices and Fields, Wiley, NJ. https://doi.org/10.1002/jnm.2516.
[10] Khursheed, A., and K. Khare, "Designing Dual-Chirality and Multi-Vt Repeaters for Performance Optimization of 32 nm Interconnects", Circuit World, Vol. 46, No. 2, PP. 71–83, 2020. https://doi.org/10.1108/CW-06-2019-0060.
[11] Khursheed, A., and K. Khare, "Optimized Buffer Insertion for Efficient Interconnects Designs", International Journal of Numerical Modelling: Electronic Networks, Devices and Fields, PP. e2748, 2020. https://doi.org/10.1002/jnm.2748.
[12] Chawla, T., M. Khosla, and B. Raj, "Optimization of Double-Gate Dual Material GeOI-Vertical TFET for VLSI Circuit Design", IEEE VLSI Circuits and Systems Letter, Vol. 6, No. 2, PP. 13–25, August 2020.
[13] Kaur, M., N. Gupta, S. Kumar, B. Raj, and Arun Kumar Singh, "RF Performance Analysis of Intercalated Graphene Nanoribbon Based Global Level Interconnects", Journal of Computational Electronics, Springer, Vol. 19, PP. 1002–1013, June 2020.
[14] Wadhwa, G., and B. Raj, "An Analytical Modeling of Charge Plasma Based Tunnel Field Effect Transistor with Impacts of Gate underlap Region", Superlattices and Microstructures, Elsevier, Vol. 142, PP. 106512, June 2020.
[15] Singh, S., and B. Raj, "Modeling and Simulation Analysis of SiGe Hetrojunction Double Gate Vertical t-shaped Tunnel FET", Superlattices and Microstructures, Elsevier, Vol. 142, PP. 106496, June 2020.
[16] Singh, S., and B. Raj, "A 2-D Analytical Surface Potential and Drain Current Modeling of Double-Gate Vertical T-Shaped Tunnel FET", Journal of Computational Electronics, Springer, Vol. 19, PP. 1154–1163, April 2020.
[17] Singh, S., S. Bala, B. Raj, and Br. Raj, "Improved Sensitivity of Dielectric Modulated Junctionless Transistor for Nanoscale Biosensor Design", Sensor Letter, ASP, Vol. 18, PP. 328–333, April 2020.
[18] Mahmoodi, H., and K. Roy, "Diode-Footed Domino: A Leakage-Tolerant High Fan-in Dynamic Circuit Design Style", IEEE Transactions on Circuits and Systems I: Regular Papers, Vol. 51, No. 3, PP. 495–503, 2004.

[19] Kumar, V., S. Kumar, and B. Raj, "Design and Performance Analysis of ASIC for IoT Applications", Sensor Letter ASP, Vol. 18, PP. 31–38, January 2020.
[20] Wadhwa, G., and B. Raj, "Design and Performance Analysis of Junctionless TFET Biosensor for High Sensitivity", IEEE Nanotechnology, Vol. 18, PP. 567–574, 2019.
[21] Wadhera, T., D. Kakkar, G. Wadhwa, and B. Raj, "Recent Advances and Progress in Development of the Field Effect Transistor Biosensor: A Review", Journal of Electronic Materials, Springer, Vol. 48, No. 12, PP. 7635–7646, December 2019.
[22] Singh, S., and B. Raj, "Design and Analysis of Hetrojunction Vertical T-Shaped Tunnel Field Effect Transistor", Journal of Electronics Material, Springer, Vol. 48, No. 10, PP. 6253–6260, October 2019.
[23] Goyal, C., J.S. Ubhi, and B. Raj, "A Low Leakage CNTFET Based Inexact Full Adder for Low Power Image Processing Applications", International Journal of Circuit Theory and Applications, Wiley, Vol. 47, No. 9, PP. 1446–1458, September 2019.
[24] Allah, M.W., M.H. Anis, and M.I. Elmasry, "High Speed Dynamic Logic Circuits for Scaled-Down CMOS and MTCMOS Technologies, in Proc", IEEE Inter. Symp. Low Power Electronics Design", 2000, PP. 123–128.
[25] Taghipour, Shiva, and Rahebeh Niaraki Asli, "Aging Comparative Analysis of High-Performance FinFET and CMOS Flip-Flops", Microelectronics Reliability, Vol. 69, PP. 52–59, 2017.
[26] Zimpeck, A.L., C. Meinhardt, and R.A.L. Reis, "Impact of PVT Variability on 20 nm FinFET Standard Cells", Microelectronic Reliability, Vol. 55, PP. 1379–1383, 2015.
[27] Navaneetha, Alluri, and Kalagadda Bikshalu, "FinFET Based Comparison Analysis of Power and Delay of Adder Topologies", Materials Today: Proceedings, Vol. 46, Part 9, PP. 3723–3729, 2021.
[28] Dadoria A.k., K. Khare, T.K. Gupta, and R.P. Singh, "Ultra Low Power High Speed Domino Logic Circuit by Using FiNFET Technology", Advances in Electrical and Electronic Engineering, Vol. 14, No. 1, PP. 66–74, 2016.
[29] Ansari, M., H. Afzali-Kusha, B. Ebrahimi, Z. Navabi, A. Afzali-Kusha, and M. Pedram, "A Near-Threshold 7T SRAM Cell with High Write and Read Margins and Low Write Time for Sub-20 nm FinFET Technologies", Integration of VLSI, Vol. 50, PP. 91–106, 2015.
[30] Meinhardt, C., A.L. Zimpeck, and R.A.L. Reis, "Predictive Evaluation of Electrical Characteristics of Sub-22 nm FinFET Technologies Under Device Geometry Variations", Microelectronic Reliability, Vol. 54, PP. 2319–2324, 2014.
[31] Sharma, S.K., B. Raj, and M. Khosla, "Enhanced Photosensivity of Highly Spectrum Selective Cylindrical Gate In1-xGaxAs Nanowire MOSFET Photodetector", Modern Physics Letter-B, Vol. 33, No. 12, PP. 1950144, 2019.
[32] Singh, J., and B. Raj, "Design and Investigation of 7T2M NVSARM with Enhanced Stability and Temperature Impact on Store/Restore Energy", IEEE Transactions on Very Large Scale Integration Systems, Vol. 27, No. 6, PP. 1322–1328, June 2019.
[33] Bhardwaj, A.K., S. Gupta, B. Raj, and Amandeep Singh, "Impact of Double Gate Geometry on the Performance of Carbon Nanotube Field Effect Transistor

Structures for Low Power Digital Design", Computational and Theoretical Nanoscience, ASP, Vol. 16, PP. 1813–1820, 2019.
[34] Goyal, C., J. Subhi, and B. Raj, "Low Leakage Zero Ground Noise Nanoscale Full Adder using Source Biasing Technique", Journal of Nanoelectronics and Optoelectronics, American Scientific Publishers, Vol. 14, PP. 360–370, March 2019.
[35] Singh, A., M. Khosla, and B. Raj, "Design and Analysis of Dynamically Configurable Electrostatic Doped Carbon Nanotube Tunnel FET", Microelectronics Journal, Elesvier, Vol. 85, PP. 17–24, March 2019.
[36] Goyal, C., J.S. Ubhi, and B. Raj, "A Reliable Leakage Reduction Technique for Approximate Full Adder with Reduced Ground Bounce Noise", Journal of Mathematical Problems in Engineering, Hindawi, Vol. 2018, Article ID 3501041, PP. 16, 15 October 2018.
[37] Wadhwa, G., and B. Raj, "Label Free Detection of Biomolecules Using Charge-Plasma-Based Gate Underlap Dielectric Modulated Junctionless TFET", Journal of Electronic Materials (JEMS), Springer, Vol. 47, No. 8, PP. 4683–4693, August 2018.
[38] Wadhwa, G., and B. Raj, "Parametric Variation Analysis of Charge-Plasma-Based Dielectric Modulated JLTFET for Biosensor Application", IEEE Sensor Journal, Vol. 18, No. 15, 1 August 2018.
[39] Yadav, D., S.S. Chouhan, S.K. Vishvakarma, and B. Raj, "Application Specific Microcontroller Design for IoT Based WSN", Sensor Letter, ASP, Vol. 16, PP. 374–385, May 2018.
[40] Singh, G., R.K. Sarin, and B. Raj, "Fault-Tolerant Design and Analysis of Quantum-Dot Cellular Automata Based Circuits", IEEE/IET Circuits, Devices & Systems, Vol. 12, PP. 638—64, 2018.
[41] Singh, J., and B. Raj, "Modeling of Mean Barrier Height Levying Various Image Forces of Metal Insulator Metal Structure to Enhance the Performance of Conductive Filament Based Memristor Model", IEEE Nanotechnology, Vol. 17, No. 2, PP. 268–267, March 2018 (SCI).
[42] Jain, A., S. Sharma, and B. Raj, "Analysis of Triple Metal Surrounding Gate (TM-SG) III-V Nanowire MOSFET for Photosensing Application", Opto-Electronics Journal, Elsevier, Vol. 26, No. 2, PP. 141–148, May 2018.
[43] Jain, N., and B. Raj, "Parasitic Capacitance and Resistance Model Development and Optimization of Raised Source/Drain SOI FinFET Structure for Analog Circuit Applications", Journal of Nanoelectronics and Optoelectronins, ASP, USA, Vol. 13, PP. 531–539, April 2018.
[44] Singh, S., and B. Raj, "Analytical Modeling and Simulation Analysis of T-Shaped III-V Heterojunction Vertical T-FET", Superlattices and Microstructures, Elsevier, Vol. 147, PP. 106717, November 2020.

Chapter 8

Noise Performance of an IMPATT Diode Oscillator at Different mm-Wave Frequencies

R. Dhar, Sangeeta Jana Mukhopadhyay, V. Maheshwari, and M. Mitra

CONTENTS

8.1 INTRODUCTION

Impact ionization avalanche transit times (IMPATTs) are high-frequency generators and amplifier devices with frequencies ranging from a few to several hundred gigahertz. IMPATTs have been the most popular among the military, defense and radar applications and even in nonmilitary applications. IMPATTs have to be mounted on resonators to be used for sustainable use; otherwise, they cannot be used for very high-power applications. In double-drift-region (DDR) IMPATTS, both the holes and the electrons have different individual drift regions, which helps in more power handling and more efficient operations. The theory and development of single-drift-region (SDR) and DDR IMPATTs over the years, as presented in literature [1–5], have made it possible for the devices to be used for high power radiofrequency (RF) applications.

Although IMPATTs are very useful devices, the main drawback of IMPATTs is that they are very noisy devices, with noise measures ranging between 30–60 dB or even more, which causes their efficiency to decrease. The main source of noise in these devices is the avalanching process, which includes the collision of electrons in a very thin avalanche region, leading to the generation of electron–hole pairs, or EHPs. There are a lot of other noises too, such as the shot noise and Johnson noise, but their effect is suppressed by the avalanching noise. The avalanching region in IMPATT devices plays a very important role in determining the noise performance of the devices [6–8].

DOI: 10.1201/9781003311379-8

In this chapter, we have presented the noise performance on Si DDR IMPATTs over a range of different window frequencies and low millimeter-wave frequencies. Gummel and Blue [9] have presented a small-signal theory on the avalanche noise of these devices. They proposed a generalized small-signal model to study the avalanche noise of an IMPATT diode having any arbitrary doping profile. They considered realistic field-dependent ionization rates of charge carriers in their model but assumed that the drift velocity of electrons and holes are saturated and independent of electric field even at the edges of the depletion layer. In their simulation with realistic diodes, they have shown that IMPATT devices can achieve 20–30-dB noise measures using different current densities from 100–1000 A/cm2, parasitic resistances of 0–1 ohm, and frequencies ranging from 7–40 GHz. Hines [10] has, on the other hand, has presented the large signal noise properties along with the frequency conversion effects in IMPATT devices for the X-band frequencies. The same motivation has been used in this chapter, with much higher frequency ranges, starting from 94 GHz to nearly 400 GHz, with a constant parasitic capacitance of 1.5 ohm. Frequency up-conversion at the test frequencies can affect the noise performance of these devices. Moreover, the temperature is kept fixed at 300K, which is one of the idealities that we have considered here. Moreover, the high-noise measures in the devices can affect the sustainability of the frequency output with a considerable amount of power. Thus, studying the noise performance of these devices at these frequencies is very necessary.

Since the noise depends on the ionization of the device at the junction, the way by which the ionizations depend on the frequency of operation is an important thing to study. Sze and Ryder [11] and Banerjee and Banerjee [12] have shown that the depletion layer width varies with the following relation:

$$W = 0.37V_{sn} / f_d . \tag{8.1}$$

The relation clearly shows that the depletion width W is inversely proportional to the desired frequency of operation f_d.

8.2 SIMULATION METHODS

Giblin et al. [13] reported the computer simulation of noise characteristics of the IMPATT diodes. Also, Mathur et al. [14] and Kuvas [15] have reported on the small-signal noise analysis and intrinsic noise theories for IMPATT devices. Mathur et al. [14] used a SDR-type IMPATT device while the one that we have used in our study is a DDR type whose structure is shown in Figure 8.1. In the figure, W = Wn + Wp represents the total active region of the diode. The region denoted by x_A represents the avalanching region of the diode [17–20]. The J_0 is the total current density of the diode,

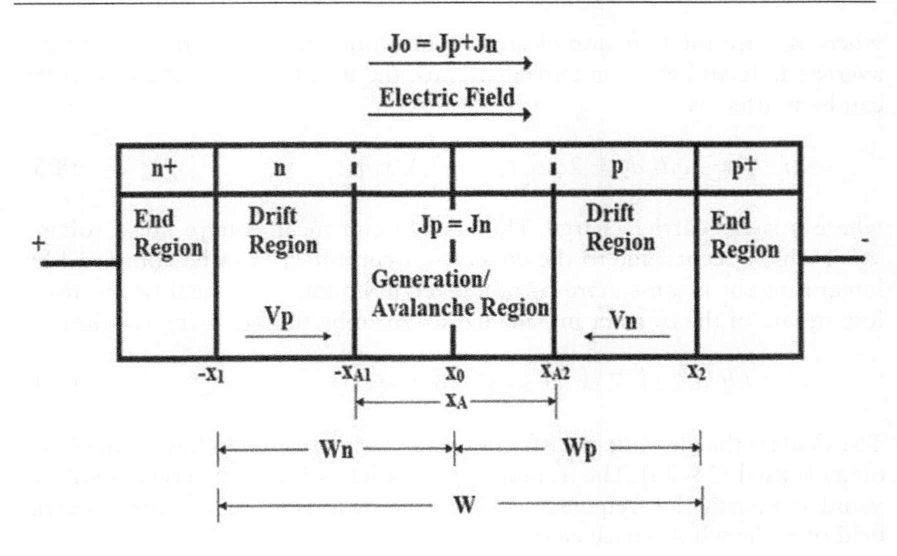

Figure 8.1 Diagram of an Si DDR IMPATT.

Table 8.1 Structural and Doping Parameters for Si for Different Operating Frequencies

Operating Frequency (GHz)	*Wn (um)*	*Wp (um)*	*ND*	*NA*	*Nn+*	*Np+*
94	0.32	0.30	1.50	1.55	5.0	2.7
140	0.28	0.245	1.80	2.10	6.2	3.5
220	0.18	0.16	3.95	4.59	7.5	4.7
300	0.132	0.112	6.00	7.32	9.0	5.9

and it flows from the N-side to the P-side. Vp and Vn represent the drift velocities of the holes and electrons, respectively. The x_0 is the center of the diode, and it is considered that the avalanching process starts at that point.

The different structural and doping parameters used for the simulation are taken from Table 8.1. The ion densities like substrate acceptor ion concentration (N_A), substrate donor ion concentration (N_D), doping concentration of n region (N_{n^+}), and doping concentration of p region (N_{p^+}) are all in the order of $10^{23}/m^3$ and $10^{26}/m^3$, respectively [21–23].

In a DDR diode, the regions are structured as$p^+p\text{-}nn^+$, and there are two separate drift regions for the holes and electrons. An elemental current di_c in the avalanche region over a small region dx is given by

$$di_c = (\alpha_p I_p + \alpha_n I_n)dx, \tag{8.2}$$

where $\alpha_{p,n}$ are the hole and electron ionization coefficients and $I_{p,n}$ are the average hole and electron currents. Thus, the mean-squared of this current can be written as

$$< di_c^2 > = 2qdi_c df = 2q(\alpha_p I_p + \alpha_n I_n)dxdf \, , \tag{8.3}$$

where q is the carrier charge. The open-circuit mean-square noise voltage $<v^2>$, which occurs due to the conversion current di_c, can be computed by integrating the mean-square current given in Equation 8.3 against the absolute square of the transfer impedance, as given by the following relation:

$$< v^2 > / df = 2q \int \left| Z_t(x, \omega) \right|^2 (\alpha_p I_p + \alpha_n I_n) dx. \tag{8.4}$$

To calculate the absolute transfer impedance $Z_t(x, \omega)$ the following methodology is used [24–26]. The terminal noise voltage due to the noise source at x and at a particular frequency ω is attained by integrating the noise electric field over the whole space charge region:

$$v_t(x, \omega) = \int_{x=0}^{x=W} e_n(x, \omega) dx. \tag{8.5}$$

The transfer impedance is measured by dividing the noise voltage by the noise current generated due to the noise source:

$$Z_t(x, \omega) = \frac{v_t(x, \omega)}{i_n(x, \omega)}. \tag{8.6}$$

Thus, by substituting Equation 8.6 into Equation 8.4, we can finally get the value of $<v^2>$.

The noise measure (M_N) is the temperature-dependent parameter and is given by the following expression:

$$M_N = \frac{\langle v_n^2 \rangle / df}{4k_B T_j(-Z_R - R_S)}, \tag{8.7}$$

where k_B is the Boltzmann's constant, T_j is the junction temperature, Z_R is the negative resistance of the device, and the R_S is the parasitic capacitance.

8.3 RESULTS AND DISCUSSION

The entire simulation has been done in a MATLAB environment with different window or operating frequencies of 94 GHz, 140 GHz, 220 GHz and 300 GHz, respectively, and the following results have been obtained. The results show that the noise measure decreases with the frequency which is shown in the following figures. The thing to notice in all the figures is that the least measure of noise is not perfectly placed at the respective window

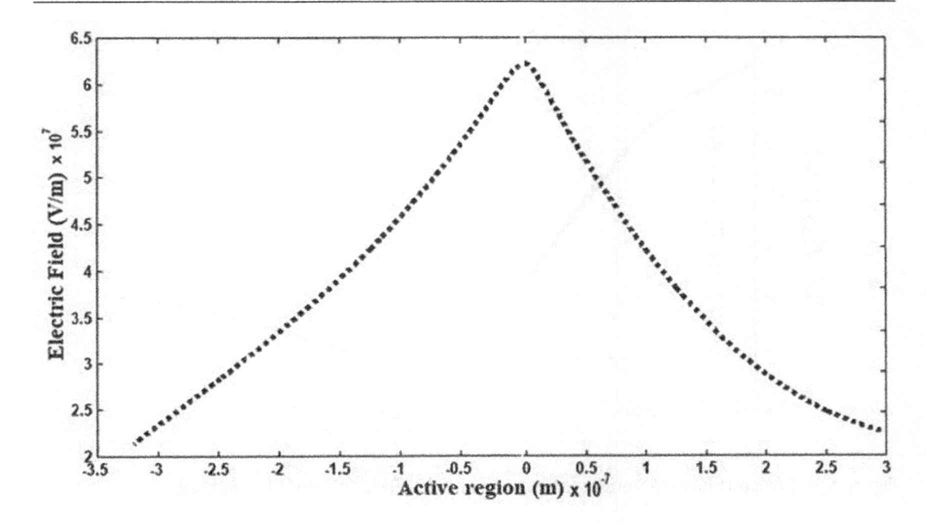

Figure 8.2 Electric field distribution for Si DDR IMPATT diode.

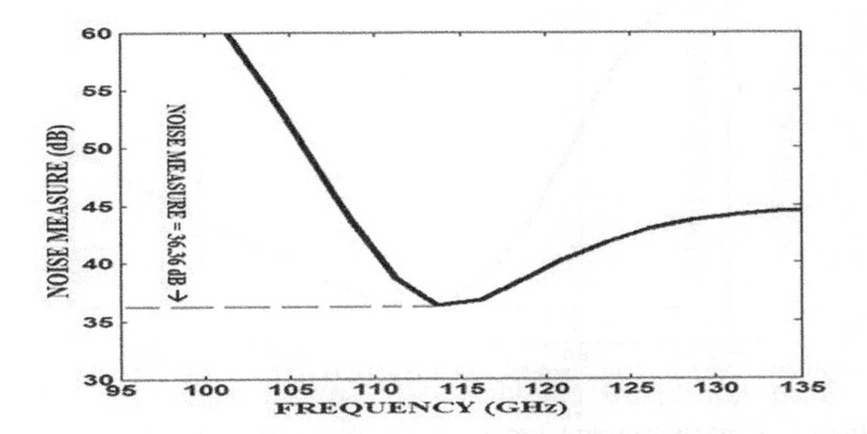

Figure 8.3 Noise measure versus frequency at 94-GHz window frequency.

frequencies but a little offset from them. The reason is justified by the non-ideality of the diodes [27–30].

Figure 8.2 shows the electric field profile at the junction for the current Si-based IMPATT device under simulation.

Figure 8.3 shows the noise measure at a window frequency of 94 GHz, and the noise measure is about 36.36 dB at a frequency of 117 GHz.

Figure 8.4 shows the noise measure at a window frequency of 140 GHz, and the noise measure is about 19.72 dB at a frequency of 218 GHz.

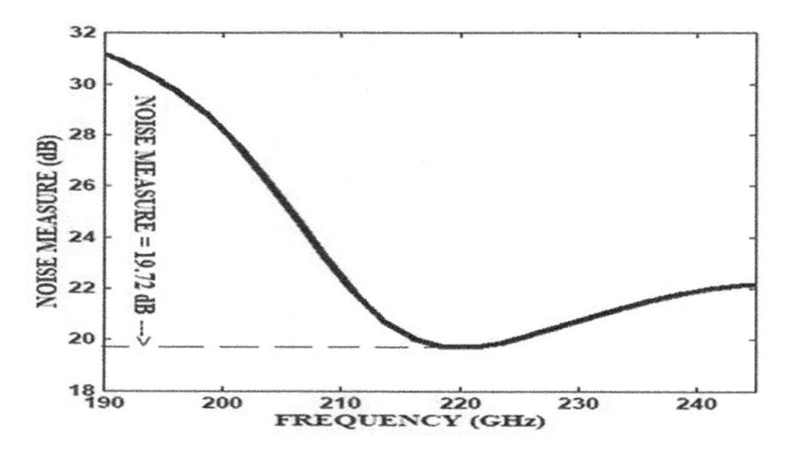

Figure 8.4 Noise measure versus frequency at 140-GHz window frequency.

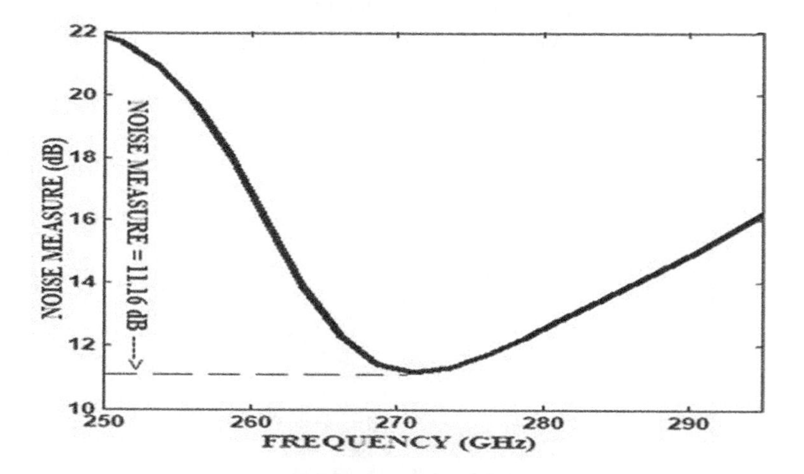

Figure 8.5 Noise measure versus frequency at 200-GHz window frequency.

Figure 8.5 shows the noise measure at a window frequency of 200 GHz, and the noise measure is about 11.16 dB at a frequency of 270 GHz.

Figure 8.6 shows the noise measure at a window frequency of 300 GHz, and the noise measure is about 0.64 dB at a frequency of 117 GHz. Thus, it can be seen that with the increase of operating frequencies the noise measure reduces significantly to less than 1 dB at about 300 GHz, which is shown in Figure 8.7.

The validation for a decrease in noise figure with frequency can be explained by the work done by Sze and Ryder [11] and Banerjee and Banerjee [12].

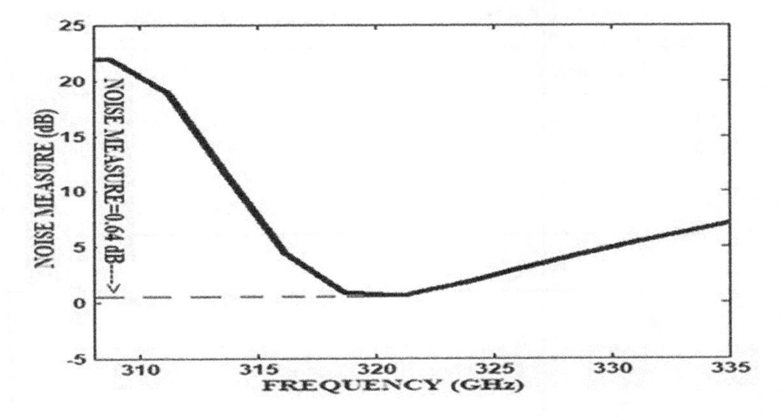

Figure 8.6 Noise measure versus frequency at 300-GHz window frequency.

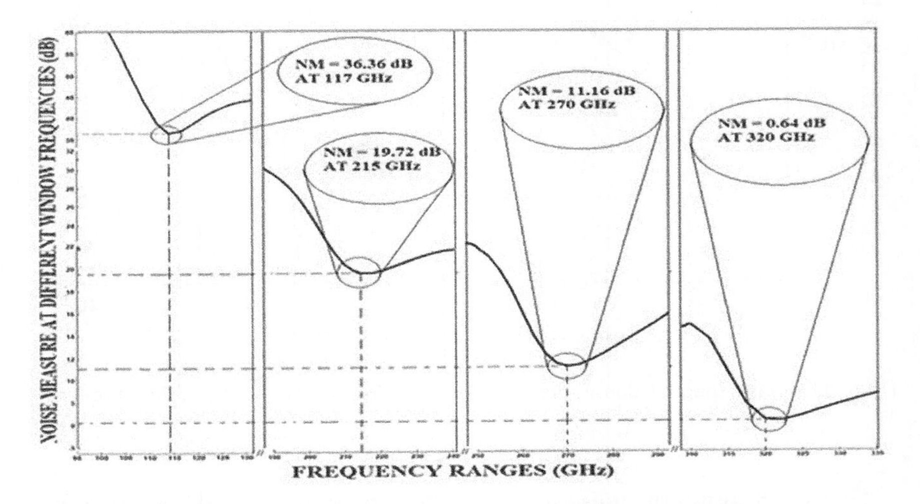

Figure 8.7 Noise measure versus frequency for progressive window frequencies.

They have shown that as we increase the operating frequency, the depletion width decreases, and hence, the avalanching process is confined to a very small region; as a result, the collision between the ions decreases, and hence, the noise measure reduces significantly [31–33]. Thus, at high frequencies, it can be concluded that the diode can be used much more efficiently, but at even higher frequencies, the depletion width will reduce so much that the ionization will cease to occur, rendering the diode to be absolutely useless as there will be no production of power at any frequency. The values of all the noise measures at the different operating frequencies are given in Table 8.2.

Table 8.2 Window Frequencies and the Respective Noise Measures

Window Frequency (GHz)	*Noise Measure (dB)*
94	36.36
140	19.72
200	11.16
300	0.64

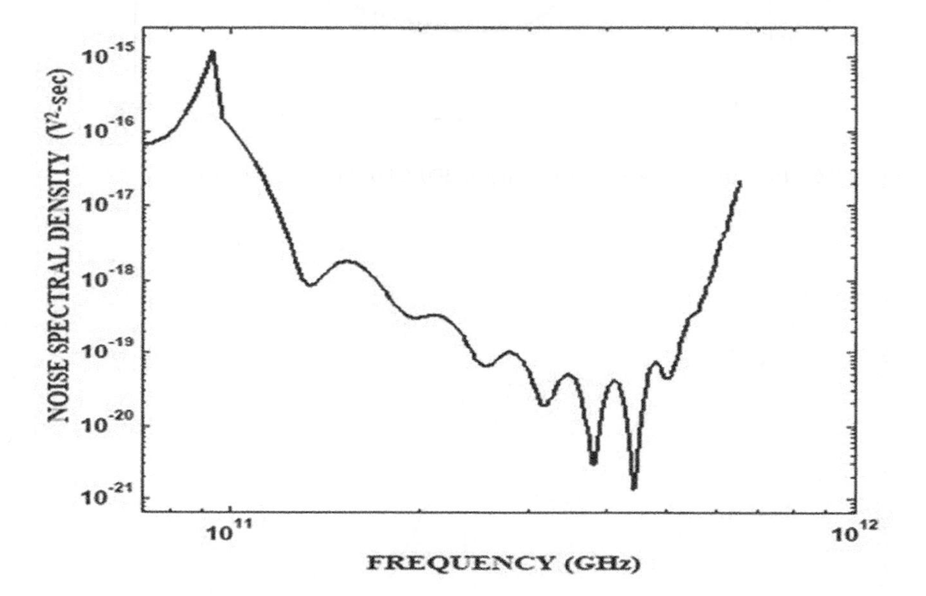

Figure 8.8 Noise spectral density versus frequency for the whole operating frequency range.

Figure 8.8 shows the noise spectral density over the desired operating frequency range. The points where the noise spectral density (NSD) takes the dip in the curve are the points of the desired frequency. The noise measure is then measured, taking the values of the NSD at those points [34–36].

The Z_R is the frequency-dependent thing in the relation. The negative resistance is obtained from the conductance–susceptance curve, or the G-B curve. The point of frequency at which the lowest conductance is achieved is considered the operating frequency. The negative resistance is obtained from the inversion of the conductance value. Hence, the lowest conductance turns into the highest negative resistance at the desired frequency. Hence, the value of the noise measure decreases with an increase in negative resistance. According to the results shown earlier, it can be assumed that an increase in frequency leads to an increase in negative resistance, which, in turn, leads to the decrease in the noise measure.

Moreover, at higher frequencies, according to Equation 8.1, we find that as the operating frequency increases, the depletion width of the diode decreases; hence, the space width for the noise to generate decreases; thus, this also acts as a method of decreasing noise measure at higher frequencies [37–40].

8.4 CONCLUSION

After obtaining all the results we can conclude that operating the IMPATT diodes at higher frequencies can help reduce the noise measures of the device, thus effectively improving the noise performance of the device. As discussed earlier, one of the reasons for the reduction in the noise measure at higher frequencies is the reduction of depletion width at high frequencies. The reduction in the depletion region also leads to a decrease in ionization and thus may affect the performance of the device. The power handling capacity of the device reduces due to decrease in ionization, which is not a favorable property for IMPATT diodes.

Thus, in order to achieve optimum performance from the device, the trade-off between power and noise must be considered carefully. Thus, to achieve a good balance between power and noise, the device must be carefully fabricated using proper structural and doping parameters required for the device to perform at that desired operating frequency.

REFERENCES

[1] Midford, T.A., and R.L. Bernick, "Millimeter -Wave CW IMPATTdiodes and Oscillators", IEEE Transactions on Microwave Theory and Techniques, Vol. 27, PP. 483–492, May 1979.

[2] Chang, Y. et al., "Millimeter-Wave IMPATT Sources for Communication Applications", IEEE MTT-S International Microwave Symposium Digest, PP. 216–219, 21–23 June 1977.

[3] Ghoshal, D., "Measurement of Electrical Resistance of Silicon Single Drift Region IMPATT Diode Based on the Study of the Device and Mounting Circuit at Threshold Condition", Journal of Electron Devices, Vol. 11, PP. 625–631, November 2011.

[4] Seidel, T.E., and D.L. Scharfetter, "High-Power Millimeter Wave IMPATT Oscillators with Both Hole and Electron Drift Spaces Made by Ion Implantation", Proceedings of the IEEE, PP. 1135–1136, July 1970.

[5] Seidel, T.E., W.C. Niehaus, and D.E. Iglesias, "Double-Drift Silicon IMPATT's at X Band", IEEE Transactions on Electron Devices, Vol. Ed-21, No. 8, PP. 523–531, August 1974.

[6] Gummel, H.K., and D.L. Scharfetter, "Avalanche Region of IMPATT Diodes", The Bell System Technical Journal, PP. 1797–1827, December 1966.

[7] Gupta, Madhu-Sudan, "Noise in Avalanche Transit-Time Devices", Proceedings of the IEEE, PP. 1674–1687, December 1971.
[8] Dallnan, G.C., and L.F. Eustman, "Avalanche-Diode Microwave Noise Generation Experiments", IEEE Transactions on Electron Devices, PP. 416, June 1968.
[9] Gummel, Hermann K., and James L. Blue, "A Small Signal Theory of Avalanche Noise in IMPATT Diodes", IEEETransactions on Electron Devices, Vol. Ed-14, No. 9, September 1967.
[10] Hines, Marion E., "Large-Signal Noise, Frequency Conversion and Parametric Instabilities in IMPATT Diode Networks", Proceedings of the IEEE, Vol. 60, No. 12, December 1912.
[11] Sze, S.M., and R.M. Ryder, "Microwave Avalanche Diodes", Proceedings of the IEEE, Vol. 59, PP. 1140–1142, 1971.
[12] Banerjee, Soumen, and J.P. Banerjee, "Effect of Punch through Factor on the Breakdown Characteristics of 4H-SiC IMPATT Diode", Proceedings of International Conference on Microwave, Jaipur, India, Vol. 08, PP. 59–62.
[13] Giblin, Roger A., Ernst F. Scherer, and Reinhard L. Wierich, "Computer Simulation o instability and Noise in High-Power Avalanche Devices", IEEE Transactions on Electron Devices, PP. 404–418, April 1973.
[14] Mathur, P.C. et al., "A Small Signal Noise Analysis for GaAs IMPATT Diode", Physica Status Solidi (a), Vol. 46, No. 321, Subject classification: 14.3.3; 22.2.1, PP. 321–326, 1978.
[15] Kuvas, Reidar L., "Noise in IMPATT Diodes: Intrinsic Properties", IEEE Transactions on Electron Devices, Vol. ED-19, No. 2, PP. 220–233, February 1972.
[16] Singh, S., and B. Raj, "Analytical Modeling and Simulation Analysis of T-Shaped III-V Heterojunction Vertical T-FET", Superlattices and Microstructures, Elsevier, Vol. 147, PP. 106717, November 2020.
[17] Chawla, T., M. Khosla, and B. Raj, "Optimization of Double-gate Dual Material GeOI-Vertical TFET for VLSI Circuit Design", IEEE VLSI Circuits and Systems Letter, Vol. 6, No. 2, PP. 13–25, August 2020.
[18] Kaur, M., N. Gupta, S. Kumar, B. Raj, and Arun Kumar Singh, "RF Performance Analysis of Intercalated Graphene Nanoribbon Based Global Level Interconnects", Journal of Computational Electronics, Springer, Vol. 19, PP. 1002–1013, June 2020.
[19] Wadhwa, G., and B. Raj, "An Analytical Modeling of Charge Plasma based Tunnel Field Effect Transistor with Impacts of Gate underlap Region", Superlattices and Microstructures, Elsevier, Vol. 142, PP. 106512, June 2020.
[20] Singh, S., and B. Raj, "Modeling and Simulation analysis of SiGe Hetrojunction Double Gate Vertical T-Shaped Tunnel FET", Superlattices and Microstructures, Elsevier, Vol. 142, PP. 106496, June 2020.
[21] Singh, S., and B. Raj, "A 2-D Analytical Surface Potential and Drain Current Modeling of Double-Gate Vertical T-Shaped Tunnel FET", Journal of Computational Electronics, Springer, Vol. 19, PP. 1154–1163, April 2020.
[22] Singh, S., S. Bala, B. Raj, and Br. Raj, "Improved Sensitivity of Dielectric Modulated Junctionless Transistor for Nanoscale Biosensor Design", Sensor Letter, ASP, Vol. 18, PP. 328–333, April 2020.

[23] Kumar, V., S. Kumar, and B Raj, "Design and Performance Analysis of ASIC for IoT Applications", Sensor Letter ASP, Vol. 18, PP. 31–38, January 2020.
[24] Wadhwa, G., and B. Raj, "Design and Performance Analysis of Junctionless TFET Biosensor for High Sensitivity", IEEE Nanotechnology, Vol. 18, PP. 567–574, 2019.
[25] Wadhera, T., D. Kakkar, G. Wadhwa, and B. Raj, "Recent Advances and Progress in Development of the Field Effect Transistor Biosensor: A Review", Journal of Electronic Materials, Springer, Vol. 48, No. 12, PP. 7635–7646, December 2019.
[26] Singh, S., and B. Raj, "Design and Analysis of Heterojunction Vertical T-Shaped Tunnel Field Effect Transistor", Journal of Electronics Material, Springer, Vol. 48, No. 10, PP. 6253–6260, October 2019.
[27] Goyal, C., J.S. Ubhi, and B. Raj, "A Low Leakage CNTFET Based Inexact Full Adder for Low Power Image Processing Applications", International Journal of Circuit Theory and Applications, Wiley, Vol. 47, No. 9, PP. 1446–1458, September 2019.
[28] Sharma, S.K., B. Raj, and M. Khosla, "Enhanced Photosensitivity of Highly Spectrum Selective Cylindrical Gate In1-xGaxAs Nanowire MOSFET Photodetector", Modern Physics Letter-B, Vol. 33, No. 12, PP. 1950144, 2019.
[29] Singh, J., and B. Raj, "Design and Investigation of 7T2M NVSARM with Enhanced Stability and Temperature Impact on Store/Restore Energy", IEEE Transactions on Very Large Scale Integration Systems, Vol. 27, No. 6, PP. 1322–1328, June 2019.
[30] Bhardwaj, A.K., S. Gupta, B. Raj, and Amandeep Singh, "Impact of Double Gate Geometry on the Performance of Carbon Nanotube Field Effect Transistor Structures for Low Power Digital Design", Computational and Theoretical Nanoscience, ASP, Vol. 16, PP. 1813–1820, 2019.
[31] Goyal, C., J. Subhi, and B. Raj, "Low Leakage Zero Ground Noise Nanoscale Full Adder Using Source Biasing Technique", Journal of Nanoelectronics and Optoelectronics, American Scientific Publishers, Vol. 14, PP. 360–370, March 2019.
[32] Singh, A., M. Khosla, and B. Raj, "Design and Analysis of Dynamically Configurable Electrostatic Doped Carbon Nanotube Tunnel FET", Microelectronics Journal, Elsevier, Vol. 85, PP. 17–24, March 2019.
[33] Goyal, C., J.S. Ubhi, and B. Raj, "A Reliable Leakage Reduction Technique for Approximate Full Adder with Reduced Ground Bounce Noise", Journal of Mathematical Problems in Engineering, Hindawi, Vol. 2018, Article ID 3501041, PP. 16, 15 October 2018.
[34] Wadhwa, G., and B. Raj, "Label Free Detection of Biomolecules Using Charge-Plasma-Based Gate Underlap Dielectric Modulated Junctionless TFET", Journal of Electronic Materials (JEMS), Springer, Vol. 47, No. 8, PP. 4683–4693, August 2018.
[35] Wadhwa, G., and B Raj, "Parametric Variation Analysis of Charge-Plasma-Based Dielectric Modulated JLTFET for Biosensor Application", IEEE Sensor Journal, Vol. 18, No. 15, 1 August 2018.
[36] Yadav, D., S.S. Chouhan, S.K. Vishvakarma, and B. Raj, "Application Specific Microcontroller Design for IoT based WSN", Sensor Letter, ASP, Vol. 16, PP. 374–385, May 2018.

[37] Singh, G., R.K. Sarin, and B. Raj, "Fault-Tolerant Design and Analysis of Quantum-Dot Cellular Automata Based Circuits", IEEE/IET Circuits, Devices & Systems, Vol. 12, PP. 638–64, 2018.
[38] Singh, J., and B. Raj, "Modeling of Mean Barrier Height Levying Various Image Forces of Metal Insulator Metal Structure to Enhance the Performance of Conductive Filament Based Memristor Model", IEEE Nanotechnology, Vol. 17, No. 2, PP. 268–267, March 2018 (SCI).
[39] Jain, A., S. Sharma, and B. Raj, "Analysis of Triple Metal Surrounding Gate (TM-SG) III-V Nanowire MOSFET for Photosensing Application", Opto-Electronics Journal, Elsevier, Vol. 26, No. 2, PP. 141–148, May 2018.
[40] Jain, N., and B. Raj, "Parasitic Capacitance and Resistance Model Development and Optimization of Raised Source/Drain SOI FinFET Structure for Analog Circuit Applications", Journal of Nanoelectronics and Optoelectronics, ASP, USA, Vol. 13, PP. 531–539, April 2018.

Chapter 9

Testing of Semiconductor Scaled Devices

Manisha Bharti and Tanvika Garg

CONTENTS

DOI: 10.1201/9781003311379-9

9.1 INTRODUCTION

Biosensor can be defined as a device that makes use of certain biochemical reactions that are mediated by immune systems, tissues, isolated enzymes, organelles or whole cells in order to detect chemical compounds by means of electrical, thermal or optical signals. In the field of research and development, biosensors are being extensively studied as they are easy, of low cost, rapid, highly selective and highly sensitive. They play a major role in the advancement of individualized medicine.

The most important factor for those who work with or around radiation is being aware of the radiation level around them. This could be accomplished with the use of radiation detectors. In order to perform the required task, a basic understanding of the various types of detectors is essential.

MEMS is a technology used to make tiny ICs. These ICs are a combination of mechanical and electrical components. Their size is in the range of a few micrometers to millimeters. MEMSs enable the development and production of new consumer and industrial products.

9.2 BIOSENSORS

The research and development of biosensors have led to the use of biosensors as a tool in the environmental, food, pharmaceutical and medical fields. The aim of designing biosensors is the production of a digital electronic signal. The production of a digital electronic signal is proportional to biochemical's concentration. This happens when many interfering species are present. The "biosensors" are called so because bio-functionalities, that is, recognition and catalysis, are used.[1] The biosensor's architecture includes transducers and biological components. Biosensors are devices that allow for the combination of a biological component for the detection of a physicochemical component and an analyte. This leads to the production of a signal that can be measured.

One of the earliest biosensors was introduced in 1962. This biosensor facilitated the monitoring of blood-gas levels at the time of surgery. Today, the most commonly used biosensors are glucose detectors and home pregnancy tests. A blood glucose biosensor is a biosensor that is commercially used in measuring blood glucose levels. The most regularly used biosensor applications are detecting contaminants in soil and water, toxicants and drugs in food, toxins and vitamins, and trace gas in mines. Recently, DNA-based biosensors are being used in clinical assessment as molecular diagnostic. In addition to this, biosensors are used in metastasis and the detection of ozone layer depletion. The production of new biosensors is taking place in various applications, such as DNA testing, food analysis, and drug detection. One of the famous biosensors is amperometric enzyme biosensor, which is composed of an enzyme (as a biological substance) and an electrode (as a transducer). The reason for the simplicity of the measurement circuit's

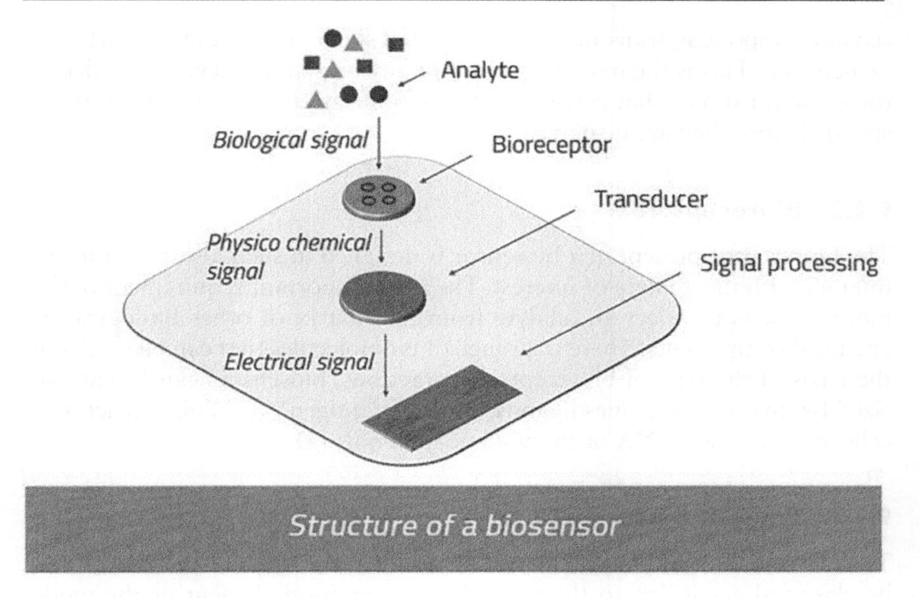

Figure 9.1 Structure of biosensor [1].

design and higher sensitivity is that the output signal is current, compared to potentiometric biosensors. The strength of the combining of enzymes to the surface of the electrodes plays a major role in the development of the amperometric enzyme biosensors. The combination of the components mentioned earlier determines how the enzyme biosensors perform.[2] To give more explanation, the dynamic range and sensitivity are determined by the electronic signal's efficiency because of enzymatic detection transferred to the electrode. This technique is known as "enzyme immobilization."

The classification of biosensors includes three generations based on how separate components have been integrated. This means that how biorecognition or bioreceptor molecule has been attached to the base transducer element. A discriminating membrane, for example, a dialysis membrane, physically traps the bioreceptor in the vicinity of the base sensor in the case of the first generation. The individual components are different from each other in the second generation, for example, electrode, control electronics and biomolecule. In the case of the third generation, the bioreceptor molecule is a part of the base-sensing element.

9.2.1 Biosensors: Design and Operations

Every biosensor functionally consists of three components. The initial part of the biosensor is the biological element that carries out the detection of the analyte and generates a signal in response. The transducer, which is the second component, transforms the generated signal into a response that can be detected. This is the most important of any biosensing device. The detector is the third part that carries out the amplification and processing of the signals before they are displayed.

9.2.2 Bioreceptors

The bioreceptor present in a biosensor is designed in such a way that it can interact with the analyte of interest. The most important requirement of the bioreceptor is to select an analyte from the matrix of other biological or chemical components. There is a range of biomolecules that can be used. On the basis of the type of bioreceptor interactions, biosensors can be categorized by involving enzymes/ligands, antibody/antigen [3] cellular structures/cells, nucleic acids/DNA or biomimetic materials [4, 5].

9.2.3 Types of Biosensors

The classification of biosensors can be done on the basis of biosensors can be classified according to the type of biorecognition element or the mode of physicochemical transduction. On the basis of the type of transducer, biosensors can be categorized as optical, electrochemical, piezoelectric and thermal biosensors.

The further classification of electrochemical biosensors can be done as amperometric biosensors (used for the measurement of the current produced at the time of reduction or oxidation of electroactive reactant or product), *potentiometric* biosensors (used for the measurement of the biosensor electrode's potential with respect to a reference electrode) and conductometric biosensors (used for the measurement of the change in conductance that arises due to the biochemical reaction).

A certain amount of light is absorbed or emitted as a result of a biochemical reaction. Optical biosensors depend on the measurement of such light. There are various optical techniques on which these optical biosensors are based on, for example, fluorescence, absorption, luminescence, *surface-plasmon* resonance (SPR) and others. SPR-based biosensors use noble metal films' plasmonic properties. The fiber optics technology invention has led to the development of numerous optical biosensors. The thermal changes in the biochemical recognition influence thermal biosensors.

There is the involvement of the measurement of change in mass that takes place due to biomolecular interaction (in piezoelectric biosensors). Piezoelectric crystals are used for the measurement of the change in mass. This is done by correlating it with the modification in oscillation frequency (of the piezo crystal).

9.3 RADIATION DETECTOR

The people who think of radiation detection, group themselves as "Geiger-counters". The most common radiation detector is the "Geiger Mueller (G-M) tube". Radiation detection devices can be classified by the type of detector element used or by its application. The instruments can be referred as a survey meter or an ion chamber or a frisker probe or a contamination meter.

9.3.1 First Radiation Detector

Since the early days of radiation testing done by Roentgen and Becquerel, researchers have tried to find ways to carry out the measurement and observation of radiation emitted by the materials. In earlier days, a photographic plate was used to data from radioactivity. It was placed in the vicinity/path of radioactive material or a beam. The radiation would develop spots on the plate. In 1896, Henri Becquerel demonstrated that radiations exist.

One of the earliest detectors was the electroscope. The electroscope used two gold leaves that repel each other when they are charged (because of the ionization occurring due to radiation). Thus, the sensitivity can be measured at a better level. Researchers could also measure beta- or alpha-particles. The understanding of basic principles of radiation could be developed by these early devices, for example, cloud chambers.

9.3.2 The Need for Radiation Detectors

In order to decide the type of detector to be used, it is important to know where and how it will be used. These types of instruments could be broadly categorized based on their applications: protection, measurement, and search.

Radiation measurement tasks are carried out where radioactive materials are required to be monitored. Awareness is the goal for this type of detection. It is required to be carried out to detect the presence of radiation where it is likely to occur.

Radiation protection is somewhat the same as radiation measurement applications. This can be said because usually it is found in a setting where there is an expectation of radiation. The goal of radiation protection is that of monitoring people. One of the most common examples is radiation dosimetry, in which radiation badges are worn by nuclear industry workers, medical personnel and others.

9.3.3 Types of Radiation Detectors

Considering radiation detection instruments, these detectors can be classified into gas-filled detectors, scintillators and solid-state detectors. All of them have their own strengths and weaknesses that decide their specific roles.

9.3.3.1 Gas-Filled Detectors

The most common radiation detector is the gas-filled detector. As the gas in the detector contacts radiation, a reaction takes place. The ionization of the gas occurs, and the electronic charge produced is measured by a meter. The gas-filled detectors can be further classified as ionization chambers, proportional counters and G-M tubes. The voltage applied across the detector is the factor that differentiates these detectors from each other.

9.3.3.2 Scintillators

Scintillation detectors are the second-most important type of detectors used in radiation detection instruments. By scintillation, it is meant to give off light. The result of a photon interacting with the scintillator material is a flash of light. The specific spectroscopic profiles are captured by scintillation detectors for the measured radioactive materials.

A scintillator material along with a photomultiplier (PM) tube is used by scintillation detectors to work through the connection of it. The pulses of light are converted into electrons by the PM tube using a photocathode material, which then amplifies that signal for the generation of a voltage pulse. This is later read and interpreted. The number of these pulses are measured over the time indicated. The strength of the radioactive source is being indicated by the pulses that have been measured.

The scintillation detectors have the ability to identify the radioactive sources. This property is useful in various radiation security applications. There are various radiation security applications, such as handheld devices, used to identify hidden radioactive material and used to monitor large populations or areas to make out differences between medical and natural sources of radiation and the sources requiring immediate concerns, for example, special nuclear material (SNM).

9.3.3.3 Solid-State Detectors

The solid-state detectors are the last important technology utilized as radiation detection instruments. Silicon solid-state detectors consist of both n-types and p-types. The depletion zone in these devices are similar to the detection area in an ion chamber. Radiation inside the depletion region interacts with the atoms and causes them to re-ionize. This creates an electronic pulse that can be measured. The detector and the depletion region are of a small scale, which means that the pairs of ions are collected very quickly. This shows that instruments using these detectors have quick response times.

This property, along with their small size, makes these detectors important in electronic dosimetry applications.

9.4 MEMSs

MEMSs are technology in which elements are miniaturized and electromechanical. They are made by taking into account microfabrication techniques. There are various types of MEMSs: Some may be simple structures without any moving elements, or others could be complex electromechanical devices with a variety of moving elements. These moving-element devices are controlled by integrated microelectronics. The most important elements are the micro-actuators and microsensors, as well as a various of elements that have a variety of functions. The category of "transducers" involves micro-actuators and microsensors. Energy is converted from one form to another by transducers. Mechanical energy is converted into an electrical energy by microsensors.

Many microsensors, including temperature sensors, pressure sensors, radiation sensors and others, have been worked on by researchers when dealing with MEMSs over several past decades. These micro-machined sensors performed better than their macro-scale counterparts. The performance of MEMS devices is not only exceptional, but they can also have low per device production costs and various other benefits. Recently, the research and development of MEMSs have come up with a number of micro-actuators, for example, micro-valves for liquid flows and control of gas; mirrors and optical switches to modulate or redirect light beams; independently controlled micro-mirrorarrays to carry out displays; micro-resonators for various applications; micro-pumps to create positive fluid pressures; and many others.

MEMSs could be formed if these miniature actuators, sensors, and structures are combined and placed on a common substrate made of silicon. The fabrication of these components is done using various micromachining processes. The merging of MEMSs with technologies other than MEMSs, such as nanotechnology, photonics and others, is called "heterogeneous integration".

A more complex integration is the future for MEMSs. The present technology for MEMSs is modest, involving a microsensor, an electronics-integrated microsensor, a micro-actuator, an electronics-integrated micro-actuator, multiple similar microsensors or multiple similar micro-actuators.

The integration of micro-actuators, microsensors and microelectronics, among other technologies, on one microchip will be the most important breakthrough in the future. This will lead to the use of various smart products because the microelectronics used to make these products have a high computational capability. MEMSs can be thought of as being the decision-making capability of this system. By thoroughly measuring the phenomena related to mechanics, heat, biology, chemistry, optics and magnetism, sensors can gather information about the environment. The processing of the information gathered by the sensors is done by the electronics, which then direct the actuators to give a response by moving, regulating, positioning, filtering and pumping. This controls the environment for some purpose or outcome. The manufacturing of MEMS devices is carried out by utilizing batch fabrication techniques. Considering the fertile and diverse nature of MEMS technology, its application areas are defined, as well as the design and manufacturing are also decided. Already, a revolutionizing effort has been made by enabling complete systems-on-a-chip in the case of MEMS.

Nanotechnology is the technology in which the manipulation of matter is carried out at the molecular or atomic level in order to make useful products at the nano-dimensional scale. Implementation could be done using two approaches, that is, the top-down and the bottom-up approach. In the top-down approach, devices and structures are formed utilizing various techniques which are similar to those used in the formation of MEMS except that they are small in size, usually by the employment of more advanced etching and photolithography methods. The technologies involving growing, deposition, or self-assembly are used in the bottom-up approach. Nano-dimensional devices have an advantage over MEMSs; that is, they have various benefits related to the laws of scaling.

It has been believed by some experts that nanotechnology

(a) allows us to put an atom or molecule in the desired place and position,
(b) allows us to make a structure or material according to the laws of physics specified at the molecular or atomic level and
(c) allows us to have the cost of manufacturing not exceeding the cost of energy and the required raw materials utilized in fabrication.

Nanotechnology and MEMSs are considered separate technologies, but they are not that distinct. They are interdependent. The scanning tunneling microscope (STM), an instrument used to detect individual molecules and atoms at the nanometer scale, is actually an MEMS device.

9.5 TESTING OF MEMSs AND SENSORS

We have to sometimes shake and bake when we are dealing with the testing of MEMSs and sensors. There is difference between standard ICs and MEMSs as well as sensors. A specific stimulus is required to get the desired testing results. So these are mostly digital. Contrary to this, MEMSs and sensors involve analog signals and circuitry. Most of the automatic test equipment provide a digital or mixed mode of testing. However, MEMSs and sensors require specialized equipment for their testing.

9.5.1 Different Techniques to Test MEMSs

Researchers have come up with numerous techniques to test MEMS devices. But none of the techniques cover all the aspects. Every technique has its own strengths and weaknesses. The following sections discuss some of these techniques that employ MEMS testing.

9.5.1.1 Atomic Force Microscopy

Atomic force microscopy (AFM), or scanning force microscopy (SFM), is a technique that has a very high resolution. It is a type of scanning probe microscopy (SPM). It provides resolution on the order of fractions of a nanometer, which is more than 1000 times more efficient than the optical diffraction limit. The AFM has a cantilever that has a sharp tip (probe) at its end. This is utilized to scan the surface of the specimen. The cantilever is actually silicon or silicon nitride whose tip radius of curvature is on the order of nanometers. When the tip is brought a sample surface, the tip and the sample experience a force between them, causing a deflection of the cantilever as per the Hooke's law [9]. The forces that are measured in AFM include van der Waals forces, mechanical contact forces, chemical bonding, capillary forces, magnetic forces, electrostatic forces, solvation forces, Casimir forces, and more. Other quantities, along with force, can be measured simultaneously by using special types of probes. Along with force, additional quantities may simultaneously be measured by using specialized types of probes.

9.5.1.2 Confocal Microscopy

Confocal microscopy or confocal laser scanning microscopy (CLSM) or laser confocal scanning microscopy (LCSM), is a technique related to optical imaging that provides an increase in the optical resolution and the micrograph's contrast by using a spatial pinhole to obstruct out-of-focus light in image formation [5]. In this process, multiple two-dimensional images at different depths of a sample are captured. This process provides the reconstruction of three-dimensional structures within a single object. This technique is mainly utilized in the industrial and scientific communities. The major applications are in life sciences, materials sciences and semiconductor inspections.

9.5.1.3 Digital Holographic Microscopy

Digital holographic microscopy (DHM) is a type of digital holography that is applied to microscopy. There is a difference between digital holographic microscopy and other microscopy methods as it does not record the object's projected image. The information embedded in the light wave that originates from the object is recorded in the form of a hologram; using this information, a numerical reconstruction algorithm is calculated by the computer. A computer algorithm is used to replace the image-forming lens. Other than these methods, related microscopies are optical coherence tomography, diffraction phase microscopy and interferometric microscopy. There is one thing common to all these methods: that they use a reference wave front to obtain phase information and amplitude (intensity). A digital image sensor or a photodetector records the information from which a computer re-creates an image of the object.

9.5.1.4 Optical Microscope

The optical microscope, also known as a light microscope, is a microscope that utilizes visible light and a system composed of lenses to produce magnified images of tiny objects. Optical microscopes are one of the oldest microscopes and were known to be invented in the 17th century. Optical microscopes are basically very simple. However, there are many complex designs that aim to enhance resolution and contrast.

The object that is needed to be viewed is positioned on a stage and is directly seen through eyepieces of the microscope. In a stereo microscope, distinct images are utilized to create a three-dimensional effect. This is in contrast with high-power microscopes in which both the eyepieces are used to view the same image. The image (micrograph) is typically captured by a camera.

There are a variety of ways in which the sample can be lit. Transparent objects can be lit from below, and solid objects can be lit with light coming through (bright field) or around the objective lens (dark field). The crystal orientation of metallic objects can be determined by using polarized light.

There are alternate ways of using optical microscopy that do not utilize visible light, including scanning electron microscopy, SPM and transmission electron microscopy. These can achieve higher order magnifications.

9.5.1.5 Scanning Electron Microscope

A scanning electron microscope (SEM) is a type of electron microscope that scans the surface using a focused electron beam and produces images of a sample. There is an interaction of electrons with atoms that produces numerous signals. These signals have information regarding the composition of the sample and the surface topography. A raster scan pattern is produced by the electron beam. The beam's position is combined with the detected signal's intensity and an image is produced. A secondary electron detector (Everhart–Thornley detector) is used to detect secondary electrons that are emitted due to excitation by the electron beam. Better resolution can be achieved by SEM which is of the order of 1 nanometer.

In conventional SEMs, specimens are seen in high vacuum or wet conditions in variable pressure or low vacuum or environmental SEMs and at a range of elevated or cryogenic temperatures with specialized instruments [6].

9.5.2 A Suitable Tester Architecture Required for Calibration and Testing of MEMSs

There is the involvement of both electrical and physical simulations in the process of MEMS testing; for example, electrical values are provided by MEMS accelerometers on the basis of applied acceleration. A trimming or calibration procedure often precedes this process. The internal registers' content is set up with some values fed by the tester to the MEMS device during this phase. These values are decided according to the responses (electrical) with respect to well-defined physical stimulations. Current MEMS testers consist of two parts, that is, a physical stimuli part and an electrical stimuli part. These separated parts are connected using wires. These wires are stretched and twisted continuously. The electrical stimuli part is related to the tester intelligence, which includes a CPU. The trimming calculation is managed by software [1]. An enhanced architecture is proposed that is used to implement the calibration and testing using hardware. MEMS accelerometers will be considered as a case study.

In calibration, a comparison of instrument outputs is done with known reference information. Then, the coefficients needed to force the output to agree with the reference are determined [2]. When we speak about accelerometers, bias and scale factor errors are the most important deterministic elements.

The calibration process is performed by the proposed tester architecture by utilizing hardware modules present in the physical stimulation unit. There is the requirement of instantiating a serial interface module (SIM) for each OUT; also a trimming computation (TC) hardware (HW) is needed.

The control signals are provided by the control unit to all the SIMs, which manages the whole process. It takes charge of calculating every trimming value on the fly with the TCHW and later sends them to SIMs. In addition to this, it carries out the communication with the electrical stimuli unit and hence receives and sends appropriate information. Direct communication is carried out between the SIM and the OUT, hence performing the trimming process. An appropriate combination of the values stored in the trimming registers and some finite state machines is made that reproduces a sequence previously identified. The selection of the relevant outputs is done, which are later transmitted to the control unit.

As a case study, three axes of a linear accelerometer are considered. It features a digital serial peripheral interface standard output and implements a prototype of the tester. This resorts to a Field Programable Gate Array design. For 1 TC and 16 SIMs 1,978 Look Up Tables, 2,140 Flip Flops and

1 Buried Random Access Memory block of 18 Kb, working at 150 Mhz, have been used.

9.6 CONCLUSION AND FUTURE CHALLENGES

Various packaging technology that are advanced are proposed for MEMSs and chips (sensors).

The same types of challenges are observed in testing packaged MEMSs (conventional) and sensors. There is an added complexity when testing a subsystem while employing 2.5-dimensional and three-dimensional techniques related to packaging.

REFERENCES

[1] https://3dprint.com/237200/dna-biosensing-with-3d-printing/

[2] Juzgado, A., A. Solda, A. Ostric, A. Criado, G. Valenti, S. Rapino, G. Conti, G. Fracasso, F. Paolucci, and M. Prato, "Highly Sensitive Electrochemiluminescence Detection of a Prostate Cancer Biomarker", Journal of Materials Chemistry B, Vol. 5, No. 32, PP. 6681–6687, 2017. https://doi.org/10.1039/c7tb01557g. PMID 32264431.

[3] Vo-Dinh, T., and B. Cullum, "Biosensors and Biochips: Advances in Biological and Medical Diagnostics", Fresenius' Journal of Analytical Chemistry, Vol. 366, No. 6–7, PP. 540–551, 2000. https://doi.org/10.1007/s002160051549. PMID 11225766. S2CID 23807719.

[4] Valenti, G., E. Rampazzo, E. Biavardi, E. Villani, G. Fracasso, M. Marcaccio, F. Bertani, D. Ramarli, E. Dalcanale, F. Paolucci, and L. Prodi, L, "An Electrochemiluminescence Supramolecular Approach to Sarcosine Detection for Early Diagnosis of Prostatecancer", Faraday Discuss, Vol. 185, PP. 299309, 2015. https://doi.org/10.1039/c5fd00096c. PMID 26394608.

[5] Pawley, J.B. (editor), Handbook of Biological Confocal Microscopy, 3rd ed. Springer, Berlin, 2006. ISBN 0-387-25921-X.

[6] Stokes, Debbie J., Principles and Practice of Variable Pressure Environmental Scanning Electron Microscopy (VP-ESEM). John Wiley & Sons, Chichester, 2008. ISBN 978-0470758748.

[7] Ciganda, L., P. Bernardi, M. Sonza Reorda, D. Barbieri, M. Straiotto, and L. Bonaria. "A Tester Architecture Suitable for MEMS Calibration and Testing", in 2010 IEEE International Test Conference IEEE, 2010, PP. 1–1.

Chapter 10

Investigation of TFETs for Mixed-Signal and Hardware Security Applications

Chithraja Rajan and Dip Prakash Samajdar

CONTENTS

DOI: 10.1201/9781003311379-10

10.1 INTRODUCTION

In 2020, when the world went into complete lockdown, there was only uncertainty and fear everywhere due to COVID-19. Fortunately, we are living in a technological era during which every service or product is available at the end of the fingertip through a smartphone or a laptop. But it is hard to imagine if the same situation had happened at the age of the vacuum tube as the computers available at that time were composed of 17000 vacuum tubes, occupied 440 $m3$ and spent 170 kW power [1]. Therefore, from the vacuum tube in 1939 to today's lightweight, fast and multifunctional microprocessor units, the electronics industry has contributed a lot to humankind. Certainly, the sole inspiration behind this achievement is the growth of the semiconductor industry following the Moore's law [2], which predicted that the transistor density on a chip would double every 18 months. Complementary metal–oxide semiconductor (CMOS) scaling reduced dimensions 100 times, clock frequency increased 1000 times but supply voltage reduced only 10 times, which contributed to 1,000,000 times power consumption [3]. This picture becomes more fascinating in the present IoT (Internet of Things) era when everything is interconnected through sensors, mixed-signal circuits, microprocessors, data storage, energy sources and transceivers, which produce enormous amounts of data that need to be digitized and processed eagerly before transmission through a secure channel power efficiently [3]. In this regard, the sigma-delta ($\Sigma\Delta$) analog-to-digital converter (ADC; [5]) is considered a compact and high-resolution architecture that contains few CMOS analog circuits [6]. However, the speed in CMOS is inversely proportional to the channel length, and process variations reduce device linearity, which demands large gain bandwidth analog circuits, and hence, $\Sigma\Delta$ ADC suffers speed and power issues. Furthermore, CMOS-based conventional cryptography security needs a lot of memory to store keys, which invalidates its use in resource-constrained IoT devices [7]. Alternatively, physical unclonable functions (PUF; [8]) is the unique chip fingerprint generated by manufacturing process variations that avoid physical storage in IoT device. However, CMOS is an age-matured technology, and hence, PUFs can be easily cracked through various modeling and cryptanalysis attacks by well-experienced attackers. So, fast, small, power-efficient and self-defensive mixed-signals hardware security modules are the need of the times.

In this perspective, tunnel field-effect transistors (TFETs; [9]) are the prominent device solution for the modern electronics that exhibits steeper subthreshold swing (SS; 60 mV/decade) due to its exclusive working principle of band-to-band tunneling (BTBT); this phenomenon assists in the scaling down of threshold voltage, along with supply voltage, to reduce the power dissipation and enhance the switching characteristics. Additionally,

TFETs have unique characteristics, such as unidirectional current conduction, Miller capacitance effect and ambipolarity, that motivate researchers to find novel circuit applications [10]. However, TFETs suffer from low driving current, poor analog/radiofrequency (RF) performance and high ambipolarity issues [11]. Therefore, the motive of this chapter is to overcome these issues through a III–V semiconductor-based hetero-material electrically doped nanowire TFET (HM-ED-NW-TFET) and then explore their applicability in ΣΔ ADC and PUF. HM-ED-NW-TFETs have been used to implement a fast and high-gain OpAmp, which supports our motive of a high-speed and accurate ΣΔ ADC construction. Finally, a novel mixed-signal-based configurable ring oscillator (CRO) PUF using a unique combination of hetero-material hetero-dielectric reconfigurable field-effect transistors (HM-HD-RFET) has been designed. The proposed CRO PUF generates 2^N security keys as compared to the 2N obtained from conventional ring oscillator (RO) PUF. All device simulations have been carried out in TCAD Silvaco tool [12]. Furthermore, circuit implementation has been done in Cadence Virtuoso [13] using a Verilog-A-based device lookup table approach, and complex mathematical computations are calculated in Matlab [14].

10.2 POWER, PERFORMANCE AND PROTECTION CONCERNS IN THE IOT

Today, more than 30 billion IoT devices are connected to each other through sensors, mixed-signal technologies, microprocessors, data storage, energy sources and transceivers. Therefore, the universal interconnectivity of the analog and digital domains has accelerated the development of power- and performance-efficient electronic systems. These applications are producing enormous amounts of real-time data that need to be digitized and processed before transmission through a secured channel. Since the presence of fast and high-speed digital ICs in every sector, such as communication, banking, defense, medical, automobile and others, are not only enhancing the dominance of digital processing but are also craving for excellent data converters or mixed-signal circuits. This picture becomes more fascinating through the enormous growth of smart devices (e.g., phones, wearables, kitchenware, medical equipment, automobiles, etc.), which provide the feasibility for realizing consumer web services through the distributed computation of data and information over a cloud in cyberspace [15]. At the end, what all devices produce are terabytes of data that need to be stored, processed and transmitted confidentially over different mediums of communication (e.g., Wi-Fi, Bluetooth, Zigbee). Therefore, power, performance and protection are the most important design metrics for IoT computing systems today.

10.3 MIXED-SIGNAL CIRCUITS FOR THE IOT

Figure 10.1 illustrates a block diagram for signal processing in which the real-world signal obtained through a sensor has been converted to a digital signal by an ADC after primary filtering and amplification. After proper processing, the signal is reconverted to analog signal by a digital-to-analog converter (DAC) and is used to control an actuator. Undoubtedly, ADC is the most critical module of a mixed-signal system. Tremendous amounts of research are in process to create power-efficient and speedy ADC architectures [16]. In a broader perspective, ADCs can be classified as Nyquist and oversampled converters. Based on the sampling rate, Nyquist ADCs have the conversion rate as twice the input signal bandwidth whereas the oversampled converters have a much higher sampling rate and, hence, require a memory element. The large analog circuitry present in a Nyquist ADC makes it prone to nonlinearity and, hence, low accuracy. On the other hand, the mixed IC architecture of an oversampling ADC has few analog components, resulting in less static issues and high resolution. The first step toward the invention of oversampled ADC was the digitization of the audio signals using ΣΔ [17] architecture, which demands a high resolution and sampling rate. Thus, oversampling and noise shaping concepts were used for the evolution of ΣΔ ADC. In spite of highly accurate conversion, the ΣΔ ADC always needs to compromise with the power budget and speed criteria [18]. This is because the scaling down of CMOS technology eventually decreases in the supply voltage, which has contributed to dense, low-power and high-speed digital circuits but which are incapable of benefiting analog and mixed-signal circuits. Of course, the major culprits here are the short channel effects (SCEs), such as drain-induced barrier lowering (DIBL), hot carrier effects (HCEs), SS degradation and velocity saturation, which are responsible for high leakage current and, hence, low-power efficiency in data converters [19]. Other than that, the reduction in supply voltage degrades the intrinsic gain, which makes an unstable bias point in an operating region and reduces the signal-to-noise ratio (SNR; [20]). Also, the transit frequency, which decides the speed of the device, is inversely proportional to channel length, and hence, a larger bandwidth is required by the analog circuits. Moreover, the process variations below 90 nm disturb bias points and, hence, affect the linearity of the circuit. Therefore, it is high time to introduce emerging low-power transistors, such as TFETs, in analog and

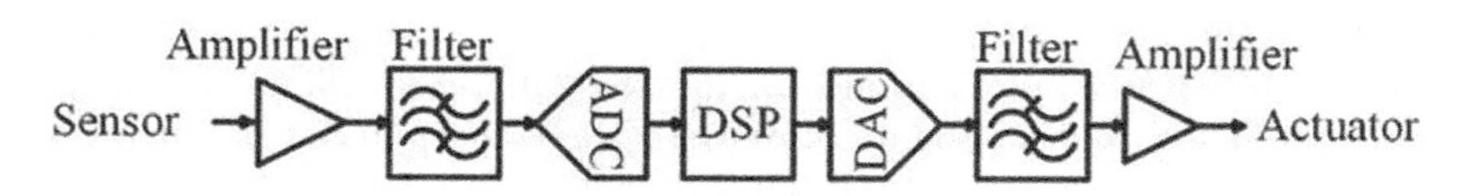

Figure 10.1 Block diagram of a basic mixed-signal system.

mixed-signal circuits, which could mitigate the aforementioned drawbacks without compromising their advantages.

10.4 HARDWARE SECURITY FOR THE IOT

When we talk about security in the IoT, it must be noted here that unlike a PC or general-purpose gadget, IoT devices are resource-constrained and dedicated in nature. Hence, the straightforward acceptance of cyber-defense solutions on the IoT platform is not as easy as it is predicted, and definitely, the nasty cyber invaders might not step out to show their brutality to these naive devices through several invasive and noninvasive attacks. Also, the software security solutions are based on the blind assumption that the physical layer is restricted from any kind of security breach. However, in actuality, the hardware module can either be a real culprit or a lifesaver. Fortunately, a number of hardware protection schemes are trending that promise to provide a high level of cyberspace security assurance to the IoT by providing safe locker, fingerprinting and camouflaging techniques [21]. In this context, it is important to highlight here that the traditional cryptographic algorithms implemented over hardware are impractical in the IoT due to its inadequate computational and secure key storage capacity. Therefore, it is high time to introduce alternative ultra-low power, fast and small self-defensive IoT modules. In this contest, PUF perfectly matches the concept of reproducing secret keys as and when required without storing them in any physical means. Alternatively, PUFs are coined as intellectual property (IP) fingerprint [22] as each IP is born with time-and space-independent uniqueness, which is random in nature and can neither be predicted nor be duplicated. However, CMOS-based PUFs are facing many challenges, such as modeling attacks (MAs), cryptanalysis attacks (CAs), side-channel attacks (SCAs), semi-invasive attacks (SIAs) and reliability issues [23], as it is an old-age, matured technology, and hence, security keys can be easily cracked through various modeling and crypto-analysis attacks by well-experienced attackers [65]. According to Rhrmair [24], a highly complex behavioral pattern is highly resistant against any machine-learning attacks and SCAs, as more training data and computational resources are needed. Again, TFETs are emerging as the most suitable alternative to PUF designs as they are well known for their unique characteristics, such as unidirectional current conduction, Miller capacitance effect and ambipolarity. Q. Huang et al. [25] experimentally found that in TFETs there is a trade-off between performance enhancement and variability suppression induced by the dominant variation source, that is, BTBT generation area, which tends to be an attractive feature for PUF designs. Furthermore, in nanoscale devices, forming an ultra-sharp doping profile, setting the fixed work function of various electrodes and obtaining exact geometric dimensions, which enhance system mismatch and encourage PUF features, are difficult tasks. Therefore, in an

IoT perspective, TFETs open up an ocean of opportunities for low-power, compact hardware security solution known as nano-TFET PUF.

10.5 LOW-POWER DESIGN TECHNIQUES

Relentless scaling of CMOSs compelled researchers to find alternative low-power chip design architectures that aim to control power performance trade-offs. Some of the existing CMOS design techniques follow:

(1) The reduced supply voltage decreases switching loss, and low V_{th} values improve device speed but results in high stand-by power loss in digital circuits. Therefore, in 1995, Mutoh and team [26] proposed the concept of multi-threshold (MT) CMOS in which low-V_{th} transistors are used to implement logic cells while the high-V_{th} FETs are used as sleep transistors. Low-V_{th} transistors are fast and dissipate high-SS leakage current whereas high-V_{th} transistors are slow and exhibit less SS leakage. The role of sleep transistors (high V_{th}) is to isolate the logic cells (low V_{th}) from power supply and, hence, control the stand-by power. However, the sudden jump of MT-CMOS from sleep to active mode drives a large current into the circuit and causes heat loss. Transistors' large wakeup latency avoids such situations. Therefore, the transistor speed and power loss can be controlled by some intermediate (drowsy) state. However, the power loss in the drowsy state would remain more, as compared to the complete sleep state.

(2) Unlike digital circuits, the power supply reduction raises many other issues in analog circuits such that when the supply voltage is reduced near to V_{th} value, the transistor overdrive voltage (V_{ov}) is limited, which causes significant variation in cut of frequency (f_t) and drain current and hence, reduce performance and power efficiency. f_t depends on carrier mobility and V_{th}. At low V_{ov}, f_t depends more on V_{th} than mobility. Therefore, the transistor bias point or V_{ov} should be optimized to reduce sensitivity [27].

(3) The ultimate power resources in the IoT are energy harvesters that are the lifelines of sensor networks. They produce few millivoltage signals that need to be digitized by high resolution and speedy ADC architectures. Therefore, ADC is the most crucial power- and performance-related module in the IoT. However, with the CMOS scaling, it is difficult to maintain optimum energy and SNR value in ADCs. In conventional successive approximation register (SAR) ADC, the capacitor count rises exponentially with resolution and noise is inversely proportional to their capacitance value, which drastically fluctuates with the layout and parasitic mismatch and, hence, deteriorates the digital output quality [28]. To mitigate these issues, noise shaping and oversampling ADCs are preferred to obtain

a low-power spectral density as oversampling allows ADCs to cope with high-speed digital processing and computation. Additionally, the noise-shaping digital filters restrict the noise signal much below original signal value, which results in a high-resolution digital output. Probably, ΣΔ is the most preferred ADC architecture for low-power and high-resolution digital conversion.

10.6 ΣΔ ADC

The general block diagram of oversampling, noise-shaping ΣΔ ADC with a digital filter and decimator are shown in Figure 10.2. The output of the ΣΔ modulator is voltage scaled to the input voltage level by a DAC in the feedback loop, which is then subtracted from the input signal. The result is then integrated and digitized using a 1-bit comparator. A suitable time-delay element at the outer comparator perfectly clocks the circuit. The integrator is the noise-shaping element, which passes the signal while keeping the noise in the high-pass band and, hence, filtering it out with the digital filter.

10.6.1 Speed and Resolution

It is evident from qualitative bandwidth and resolution trade-offs for some of the ADC techniques that ΣΔ converters attain the highest resolution for relatively low-signal bandwidths. Consequently, ΣΔ techniques are often used in speech applications in which the signal bandwidth is only 4 kHz that needs 14 bits resolution. Similarly, ΣΔ ADCs are popular for digital audio applications, in which the signal bandwidth is 20–24 kHz for which high-fidelity audio requires 16–18 bits of resolution. Flash converters, on the other hand, may be used for broadcast video applications, in which the signal band is about 5 MHz, but the resolution required is only about 8 bits. Therefore, ΣΔ converters are increasingly used for low-speed and high-resolution purposes as the digitization process is inside the feedback loop.

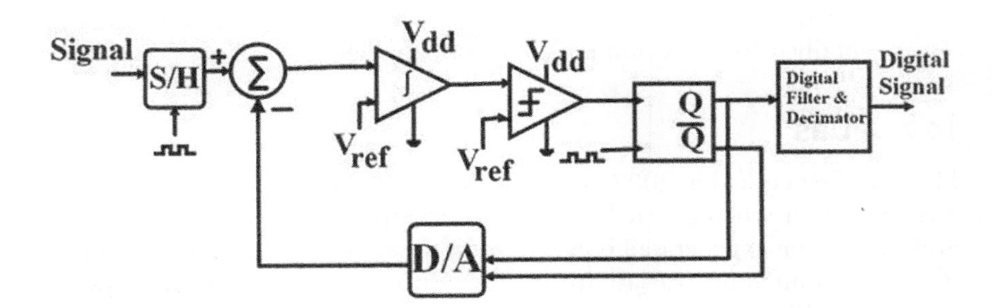

Figure 10.2 Basic ΣΔ ADC architecture.

Through this process, the circuit filters most of the self-generated noise. This resolves some of the stickier problems associated with very-high-resolution ADCs [29].

10.6.2 Linearity and Distortion

Other than SCEs, the reduced technology nodes are under a heavy load of maintaining precise doping, sharp geometric regions, fixed work functions and dielectric values because of limited fabrication methods that are likely to cause angular tilts, shadowing effect and lateral straggle [30]. Hence, these fluctuations generate nonlinearities such as harmonics, intermodulation, desensitization and gain compression issues especially in RF systems, where amplifiers and mixers usually become the victim of low-quality devices [31]. Apart from the fabrication complexities, change in environmental elements such as fluctuations, radiations and power supply variations have also become the reasons for harmonic distortions and intermodulation, which have to be investigated for better reliability of the device.

10.6.3 ADC Figure of Merits

(i) Signal-to-noise and distortion ratio (SINAD): SINAD is the ratio of the signal power to the residual signal power, which consists of both the noise and distortion components both. Generally, the distortion consists of the odd harmonic (2n+1) times input frequency components, among which the most power is associated with the first three components.
(ii) SNR: SNR compares the signal power to that of the noise power and ignores the harmonic components.
(iii) Spurious-free dynamic range (SFDR): SFDR is the difference in the signal peak power and the highest non–signal peak power.
(iv) Total harmonic distortion (THD): THD decides the linearity of the ADC. A low THD indicates that the devices do not get saturated for large signal values.
(v) Effective number of bits (ENOB): ENOB are the number of bits obtained for a DC input signal through ADC.

10.7 PUFS

PUF was first coined in 2002 and is like a mathematical function that generates an output when queried with an input. Input is known as a challenge, and the output so generated is called a response. Coincidently, due to unpredictable manufacture variability PUF produces unique challenge response pairs, or CRPs. PUFs are classified based on CRP count and variation as shown in Figure 10.3. Depending on the number of CRPs generated, PUFs

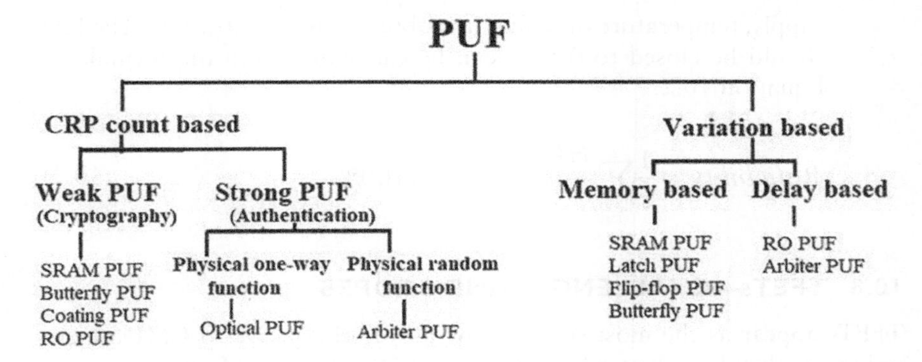

Figure 10.3 PUF classification.

are further classified as weak or strong PUFs. Weak PUFs generate few CRPs and are used in cryptography. Static Random Access Memory (SRAM), butterfly, coating and RO PUFs are known as weak PUFs. On the other hand, optical and arbiter PUFs are known as strong PUFs. Similarly, variation PUFs are further classified as memory and delay-based PUFs. SRAM, latch, flipflop and butterfly PUFs are examples of memory PUFs whereas ROs and arbiter PUFs are delay-based PUFs.

10.7.1 PUF Figure of Merits

The following are the PUF figure of merits (FoMs):

(1) Uniqueness: It determines the similarity factor among various PUFs and are calculated using inter-hamming distance (HD). In order to differentiate between various PUFs it is important to maintain an HD of at least 50%. The uniqueness can be obtained from the Equation 10.1:

$$Uniqueness = \frac{2}{j(j-1)} \sum_{i=1}^{j-1} \sum_{k=i+1}^{j} \frac{HD(R_i, R_k)}{n} \times 100\%, \quad (10.1)$$

Where Ri and Rk are n-bit response of chips i and k ($i \neq k$) and j is the number of chips.

$$HD(R_i, R_k) = \sum_{i=1}^{n} \left(r_{i,t} \oplus r_{k,t}\right)$$

(2) Reliability: It determines that the PUF response should remain consistent with changing environmental conditions, such as power

supply, temperature or aging. It is obtained using intra-HD. The HD should be closed to 0%. It can be calculated from the formula in Equation 10.2:

$$\text{Re}\,liability = \frac{1}{s}\sum_{m=1}^{s}\frac{HD\left(R_i, R'_{i,m}\right)}{n}\times 100\%. \tag{10.2}$$

10.8 TFETs—CHALLENGES AND SCOPES

TFETs appear as the most suitable low-power alternative to CMOS technology as they have better SS and reduced OFF current. However, to conquer the semiconductor industry, there are many challenges that need to be addressed in TFETs, which are directly related to their device-to-circuit applicability:

(1) Low ON current: The *ION* of a TFET depends on the tunneling probability of the BTBT process. The tunneling probability can be calculated using WKB (Wentzel–Kramers–Brillouin) approximation [32].

(2) Ambipolarity: Unlike a metal–oxide–semiconductor field-effect transistor (MOSFET), a TFET possesses an asymmetrical source and drain doping and, therefore, behaves significantly different for a positive or negative *VGS* polarity. Positive *VGS* reverse-biases the source channel regions and current flows due to band bending at their junction. Similarly, for negative *VGS*, the drain channel regions get reverse biased, and the band bending at their junction causes current flow. This is called an ambipolar current conduction or ambipolarity, which is an unwanted phenomenon for conventional electronic circuits and needs to be investigated rigorously to find its application in novel circuit applications.

(3) Random variations: Random dopant fluctuations (RDFs), geometric variations and gate work function variation (WFV) are the sources of variation in TFETs [33]. There are random variations from wafer to wafer and in dies across a wafer due to PVT (process voltage temperature) variations during manufacturing steps. Due to variations in devices, the magnitude of delay variations can be 5% or more [34]. All these sources of variation, which limit the component density of the IC, are the subjects of constant research. Nevertheless, the relative variations tend to increase with shrinking device sizes. The source of metal WFV can be characterized with randomized grains with an average grain size of varying shape and orientations. As the devices are scaled down, the formation of ultra-sharp doping profile, uniform doping and diffusion phenomenon from one region to another are major sources of RDF in TFET. The movement of dopant atoms at a

junction RDF leads to the reduction of abrupt junctions, and hence, an undesirable threshold voltage variation occurs. It is primarily because of local dopant variation in the source region as BTBT does not occur along a single vertical plane, which reduces ON current and degrades SS. The RDF and geometric variations induced by mask variations in manufactured ICs lead to inherent delay variations in TFETs, which is considered to be an asset for PUFs.

(4) Linearity: Linearity plays a vital role in any RF system, in which adjacent channel signals interfere with the original signal and deteriorate information quality. Other than SCEs, reduced technology nodes operate under a heavy load of maintaining precise doping, sharp geometric regions, fixed work functions, and dielectric values because of limited fabrication methods that are likely to cause angular tilts, shadowing effects and lateral straggle. These fluctuations generate nonlinearities such as harmonics, intermodulation, desensitization and gain compression issues, especially in RF systems, where amplifiers and mixers fall victim to low-quality devices. Apart from the fabrication complexities, changes in environmental elements such as fluctuations, radiation and power supply variations can also cause harmonic distortions and inter-modulation. So, these elements must be investigated for improvements in the reliability of these devices.

The following are some of the TFET design solutions extracted from the literature to prevent the previously mentioned device issues:

(1) Gate Oxide Thickness and Relative Permittivity of Dielectric: The reduction in gate oxide thickness increases the gate control over the band bending, and hence, both ON and ambipolar current increase. Alternatively, high-k gate dielectrics have effective oxide thickness greater than the conventional silicon dioxide thickness. K. Boucart and A. M. Ionescu [9] have demonstrated clearly that the drain current increases with an increasing dielectric constant as the gate coupling increases with the dielectric constants. Furthermore, the concept of multiple gates also increases the drain current as it helps with obtaining stringent control over the channel region and further supports the band bending, which, in turn, increases the tunneling rate. But increasing band bending increases not only the ON-state current but also the ambipolar current. To overcome this challenge, the idea of an asymmetric gate and an asymmetric dielectric is additionally incorporated into the architecture of TFETs.

(2) Highly Doped Pocket Insertion: Low ON-current of TFETs can also be increased by deposition of n+ pocket near the source junction using dopant engineering. This asymmetry in the device helps to improve ON-current as well as steepens the device characteristic.

Ritesh Jhaveri [35] explained, that a tunneling junction formed between a p+ region and a narrow fully depleted n+ layer under the gate acts as a source of electrons for the channel. It reduces the tunneling width and increases the lateral electric field, which results in improved device performance.

(3) Material Engineering: For decades, silicon led the semiconductor industry because of its advantages over any other material. The use of silicon in manufacturing TFETs will ease the TFET deployment issue in mainstream fabrication. But with silicon, TFETs suffer from low conduction current issues, which create a need to investigate new materials. For this, germanium III–V semiconductor and carbon-based nanomaterials have been used to realize TFETs. An article by Vinay Saripalli et al. [36] shows the comparison between homojunction TFET and heterojunction TFET. Heterojunction TFET can achieve higher ON-current because of the higher electric field strength due to staggered p-n heterojunction.

(4) GAA Structure: Increase in gate coverage ratio increases the electrostatic control over the channel region, which enhances the BTBT and increases the drain current. In Kim [37], a GAA structure is simulated for different gate coverage ratios, which shows the drain current characteristic with variation in gate coverage. The magnitude of drain current is proportional to the ratio of gate coverage. Therefore, the drain current is linearly dependent on the gate coverage. Also, the excellent electrostatic control of the gate by the GAA structure leads to volume inversion, the suppression of corner effects and the non-confinement of carriers near to oxide interface allowing better scaling options, robustness against SCEs and ideal SS. Additionally, GAA can be synthesized using a wide variety of materials and fabrication techniques, such as molecular beam epitaxy (MBE), chemical beam epitaxy (CBE), laser ablation and the metal organic vapor phase epitaxy (MOVPE; [38]).

(5) Dopingless TFETs: In submicron technology, physically doped TFETs (PD-TFETs) suffer from process variability because of RDFs. The literature also reveals that conventional doped TFETs are more sensitive toward RDFs as compared to MOSFETs because of the direct influence of tunneling width. Also, the transconductance and V_{th} are more affected by RDFs in PD-TFET, as reported in the literature. In the case of TFETs, the V_{th} is reported in two ways, depending on the saturation of the energy barrier width narrowing with gate and drain voltages. Furthermore, dopingless devices are the next era of TFETs, discovered in the form of charge plasma (CP; [39]), junctionless TFETs by Ghosh et al. [40] and electrically doped (ED) TFETs [41]. Additionally, vertical GAA polarity control nanowire FET is also presented by M. De Marchi et al. [42]. Both CP-TFETs and

ED-TFETs possess the advantage of dopingless substrates. ED-TFET additionally facilitates dynamic configuration among MOSFETs and TFETs, which means that just by changing the external voltages, we can operate a TFET as a MOSFET, which is better known as a reconfigurable FET (RFET).

10.9 CASE STUDY: IMPLEMENTATION OF A ΣΔ ADC USING AN HM-ED-NW TFET

In this section, we investigate the role of a $Al_xGa_{1-x}Sb/GaAs_{1-y}P_y$ based HM-ED-NWTFET [43] in ΣΔ ADC design, which exhibits better DC/RF and linearity characteristics as compared to the conventional TFET.

10.9.1 HM-ED-NW-TFET

The physical three-dimensional (3D) structure of n and p HM-ED-NW-TFETs is shown in Figure 10.4, and the dimensional properties are tabulated in Table 10.1. All the device simulations are done in Silvaco TCAD tool using universal Schottky tunneling (UST), drift-diffusion model, Fermi–Dirac statistics, SRH and Auger recombination and BTBT models. In the ED-TFET, the electrodes of ±0.8 V over the substrate create source and drain regions. In order to improve the device characteristics, a low-band-gap material is used in the source region, and a high-band-gap material is used in the drain/channel regions in n-HM-ED-NW-TFET while the reverse is the

Table 10.1 Device Structural Properties

Parameters	*n-Type*	*p-Type*
Oxide Thickness (t_{ox})	1nm	1nm
Body Thickness (D)	15nm	15nm
Channel Length (L_g)	30nm	30nm
Source Length (L_s)	30nm	30nm
Drain Length (L_d)	30nm	22.5nm
Source Spacer Length (L_{ss})	1nm	1nm
Drain Spacer Length (L_{ds})	5nm	12.5nm
Doping	(n-type) 1e17 atoms/cm3	(p-type) 1e17 atoms/cm3
Gate Work Function (ϕCG)	4.75 eV	4.6 eV
Over Source Work Function (ϕES)	5 eV	4.3 eV
Over Drain Work Function (ϕED)	5 eV	4.8 eV
Voltage Over Source (V_{es})	-0.8 V	+0.8 V
Voltage Over Drain (V_{ed})	+0.8 V	-0.8 V

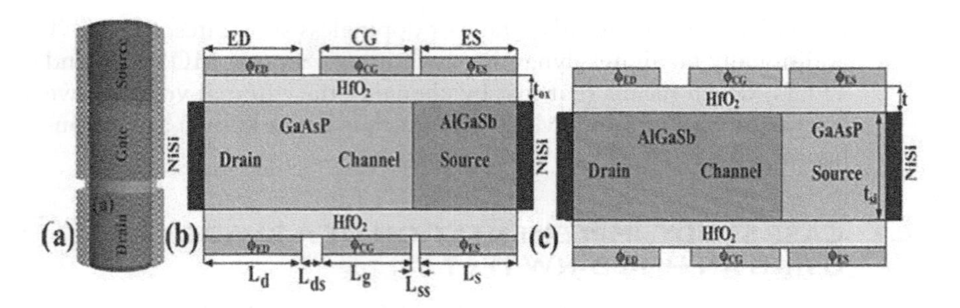

Figure 10.4 HM-ED-NW-TFET (a) three-dimensional (3D) structure, 2D structure of (b) n-type and (c) p-type.

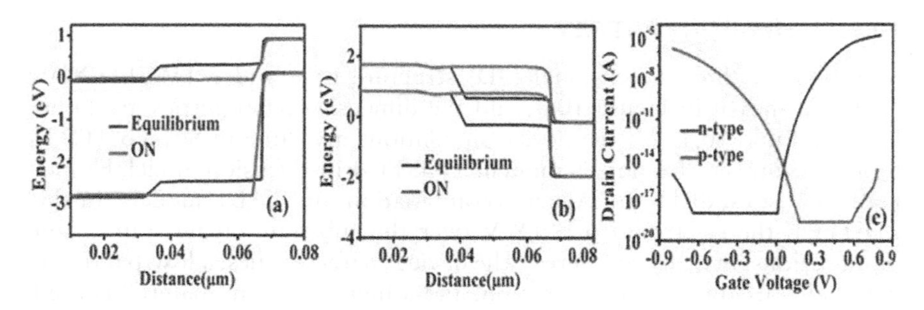

Figure 10.5 Comparison of (a) n-HM-ED-NW-TFET, (b) p-HM-ED-NW-TFET energy band diagrams and (c) drain current characteristics.

case with p-HM-ED-NW-TFET. In this proposed work, $Al_xGa_{1-x}Sb$ is the low- and $GaAs_{1-y}P_y$ is the high high-band-gap material. Figure 10.5 demonstrates the energy band diagram (EBD) of n and p-type HMED-NW-TFETs at equilibrium and ON states. During ON state when sufficient gate voltage is applied, the band bending causes tunneling of carriers from the valence band (VB) of the source toward the conduction band (CB) of the drain. The downward movement of the channel CB is caused by positive gate voltage of n-HM-ED-NW-TFET (Figure 10.5a), whereas in the case of p-HM-ED-NW-TFET (Figure 10.5b), the channel CB moves upward as negative gate voltage is applied. The drain current of the proposed n- and p-type HM-ED-NW-TFETs is shown in Figure 10.5c.

10.9.2 Implementation of ΣΔ ADC using HM-ED-NW-TFET

The basic ΣΔ ADC is constructed using HM-ED-NW-TFET and its FoMs are investigated for which a two-stage OpAmp is first constructed, and its

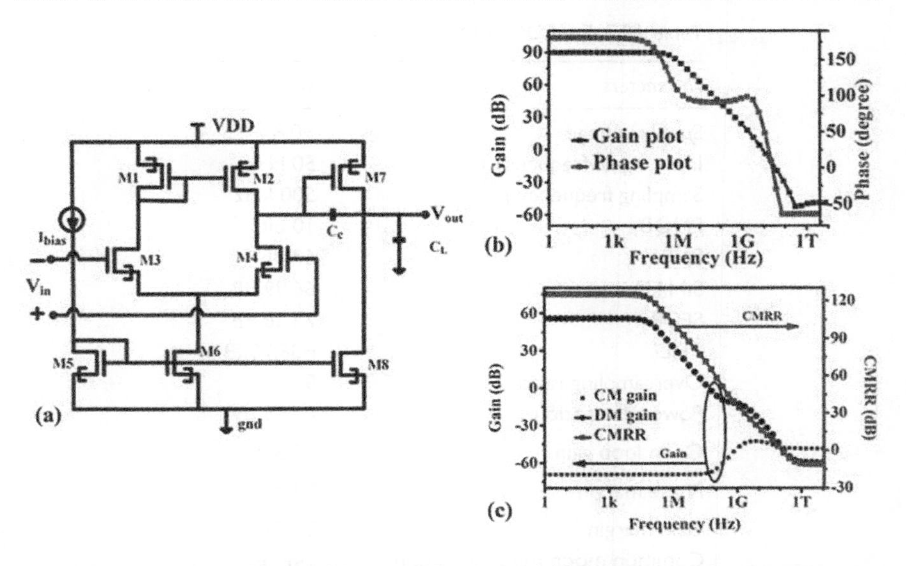

Figure 10.6 (a) Circuit diagram, (b) gain–phase plot and (c) common-mode rejection ration of HM-ED-NW-TFET-based OpAmp.

parameters are analyzed. The circuit diagram of the two-stage OpAmp consists of differential and common source amplifiers is shown in Figure 10.6a. The gain and phase plots are depicted in Figure 10.6b. The GBP of our OpAmp is 9.2 GHz, which is quite a handsome value for the successful digitization of a fast signal. The common-mode rejection ration (CMRR) is another important parameter, which decides the noise rejection capability of the the OpAmp as shown in Figure 10.6c. Finally, the output and performance metrics of ΣΔ ADC are shown in Figures 10.7a through 10.7d and Table 10.4.

10.10 CASE STUDY: IMPLEMENTATION OF A MIXED-SIGNAL-BASED CONFIGURABLE PUF USING AN RFET

During PUF authentication protocol the server saves CRPs in a table and whenever a client asks for authentication, the server sends a challenge to the client, who runs it over the PUF, and a response is sent back to the server for verification; if the CRP matches with the data in the table, then access would be given to the client. To avoid a replay attack, used CRPs are deleted from the table. Therefore, a large number of CRPs are required for long-lasting PUF usage. However, CRO-PUF produces only half (N) of the CRPs by comparing double (2N) number of Ros as shown in Figure 10.8. Additionally, with increasing number of CRPs multiplexer (MUX), counter

Table 10.2 ΣΔ ADC FoMs

Parameters	*Values*
Supply voltage	−0.8 V
Input signal frequency, f_{in}	50 kHz
Sampling frequency, f_S	500 kHz
ENOB	10 bits
SNR	67 dB
SINAD	66.99 dB
SFDR	74.38 dB
THD	−88.78 fB
Oversampling ratio	5
Power dissipation	0.325 μW
Open loop gain	89.8 dB
Phase margin	87 deg
Gain margin	21.7 dB
Common mode rejection ration	125 dB
GBP	9.2 GHz
Slew rate	34.5 V/μs
−3dB gain Bandwidth	555 kHz
Power Supply Rejection Ratio	−46.23 dB

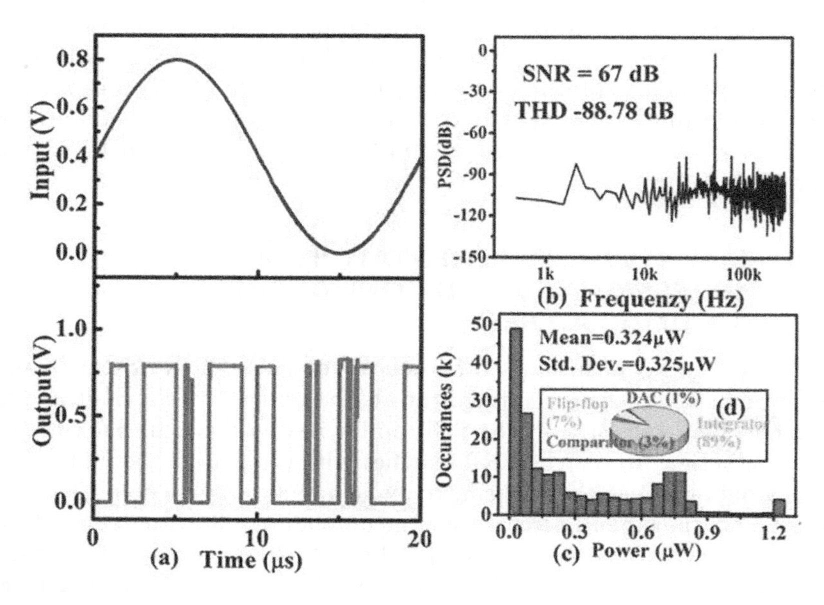

Figure 10.7 ΣΔ ADC: (a) Signal waveforms, (b) Photoshop Document, (c) power consumption and (d) distribution.

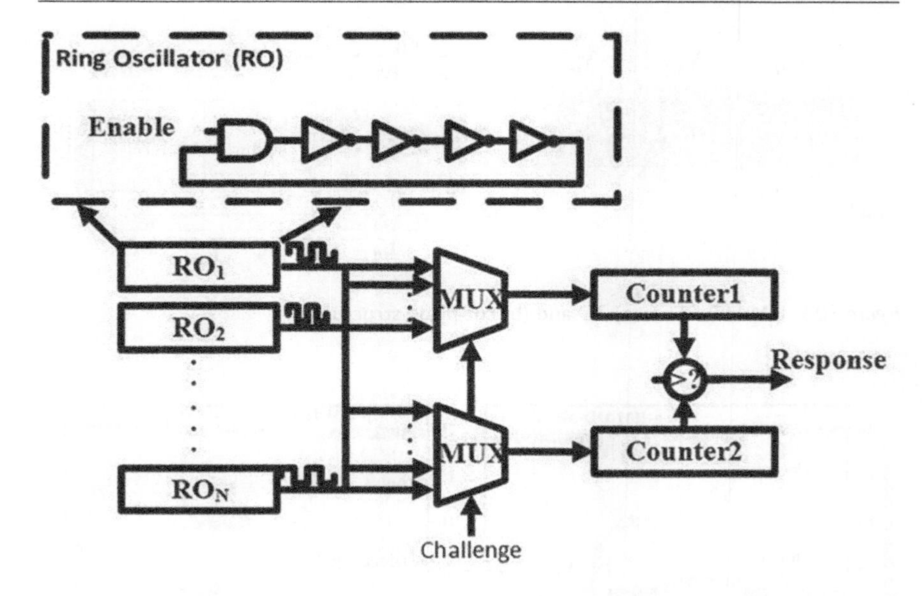

Figure 10.8 CRO architecture.

and comparator size also increase, demanding more power, space and cost. Alternatively, many researchers suggested that 2^N CRPs can be generated by inserting some configurable module (CM) between two inverters in the RO. However, they are CMOS-based architectures that are power-consuming and exposed to existing attacks. Therefore, this section presents an HM-HD-RFET based on a lightweight and low-power novel CRO-PUF architecture. Also, CRO output is digitized into a stream of response bits by a sigma-delta ADC attached at the periphery without MUX, counter and comparator.

10.10.1 Reconfigurable FET

The device structure of a HM-HD-RFET [44] is shown in Figure 10.9. The 3D and cut-plane view of the device structure is shown in Figures 10.9a and 10.9b. The device dimensions are tabulated in Table 10.3. All the device simulations are done in Silvaco TCAD tool using UST, drift-diffusion model, Fermi–Dirac statistics, SRH and Auger recombination and BTBT models. Basically, RFET is an ED device in which external electrodes over source and drain can take four possible potential values (0.8 V, 0.8 V), (0.8 V, –0.8 V), (–0.8 V. 0.8 V) and (–0.8 V, –0.8 V). Therefore, based on the potential applied on over electrodes, either electrons or holes are attracted in the source and drain regions as shown in Figure 10.10. This leads to n- or p-type MOSFET or n- or p-type TFET. Figure 10.11a shows the EBD of

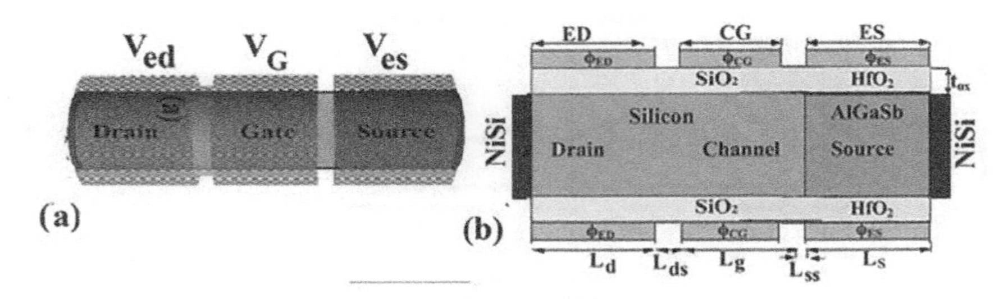

Figure 10.9 HM-HD-RFET: (a) 3D and (b) cut-plane structure.

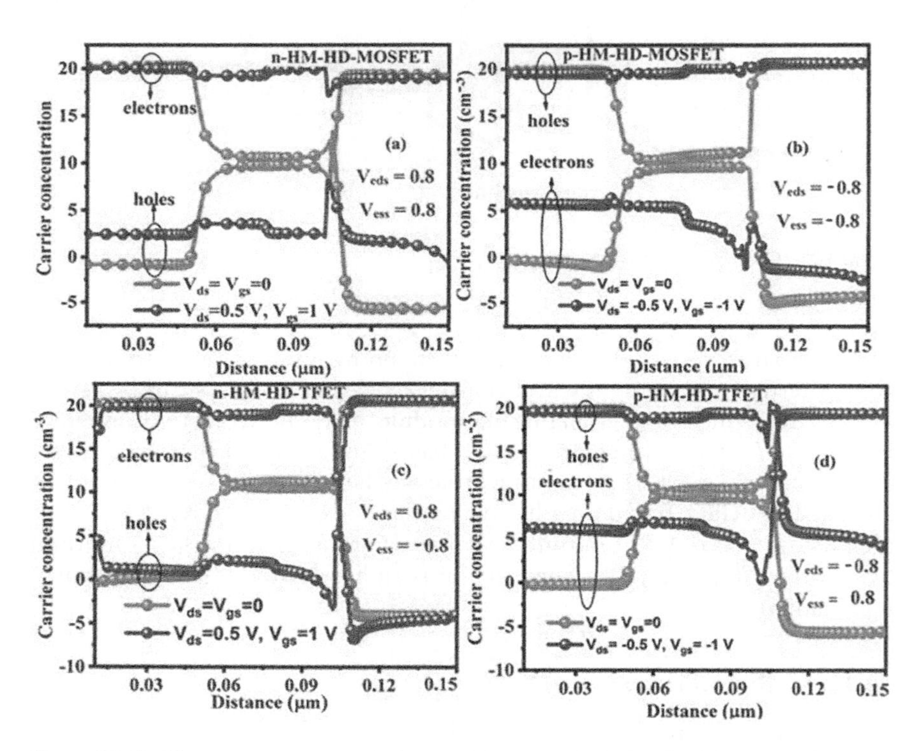

Figure 10.10 Charge carrier concentration in (a) n-HM-HD-MOSFET, (b) p-HM-HD-MOSFET, (c) n-HM-HD-TFET, and (d) p-HM-HD-TFET.

n-HM-HD-MOSFET and n-HM-HD-TFET during ON condition. It shows a clear difference between the band bending principle followed by n-type MOSFETs and TFETs in which current flows due to BTBT from source to drain region, where, as in a MOSFET, current flows by thermionic emission. Finally, Figure 10.11b shows a symmetric drain current achieved by

Table 10.3 Device Structural Properties

Parameters	*Value*
Body Thickness (D)	10 nm
Source Length (*Ls*)	50 nm
Drain Length (*Ld*)	50 nm
Channel Length (*Lg*)	50 nm
Oxide Thickness (*tox*)	2 nm
Source/Channel Space (*Lss*)	5 nm
Drain/Channel Space (*Lds*)	5 nm
Electrode Work Function	4.5 eV

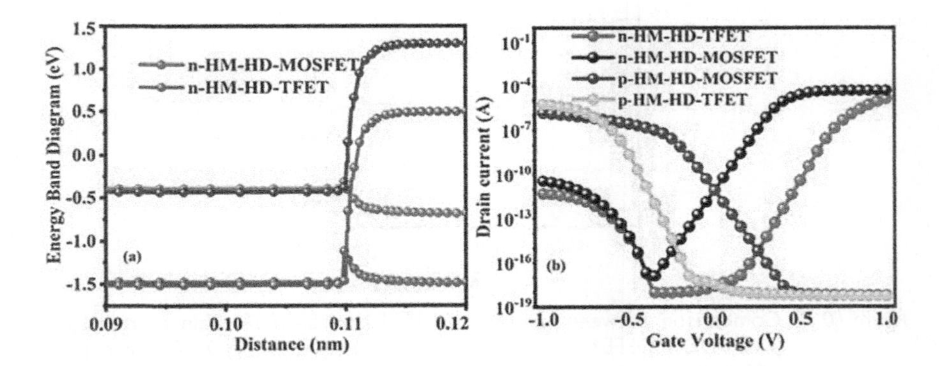

Figure 10.11 (a) EBD on n-HM-HD-MOSFET and n-HM-HD-TFET and (b) drain current characteristics of HM-HD-RFET.

four FETs, which is sufficiently high enough to drive circuit applications. Figure 10.12a compares proposed RFET electrical characteristics to that the Si-RFET, and it shows that Si-MOSFET provides better ON current as compared to HM-HD-MOSFET, but there is much improvement in the HM-HD-TFET ON current as compared to the Si-TFET, which confirms the utilization of configurable nature in a novel PUF architecture.

10.10.2 HM-HD-RFET Based Novel CRO-PUF

Figure 10.13 shows a circuit diagram of a novel CRO PUF developed using HM-HD-RFET. Here, CM is inserted between two inverters. In a CM, two HM-HD-RFET are wired in such a manner that either device works in one of the FET configurations and is in the ON condition as explained by the truth table (Table 10.4). The CRO output is then applied to a ΣΔ ADC, which produces a sequence of response bits. Finally, the CRO PUF FoMs

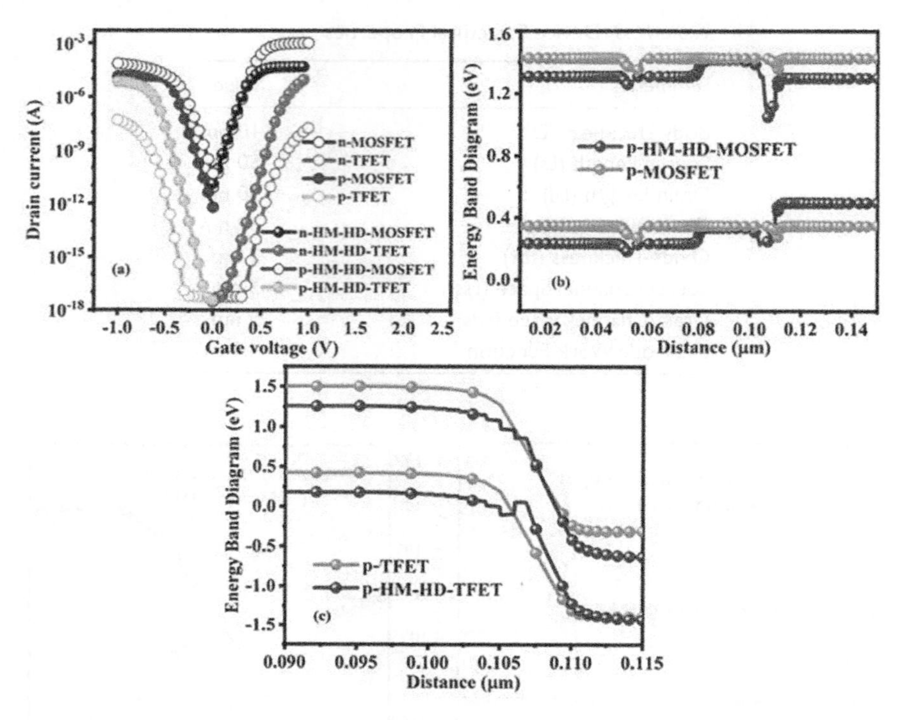

Figure 10.12 Comparison between (a) HM-HD-RFET and Si-RFET drain currents, (b) p-HM-HD-MOSFET and p-Si-MOSFET, and (c) p-HM-HD-TFET and p-Si-TFET EBDs.

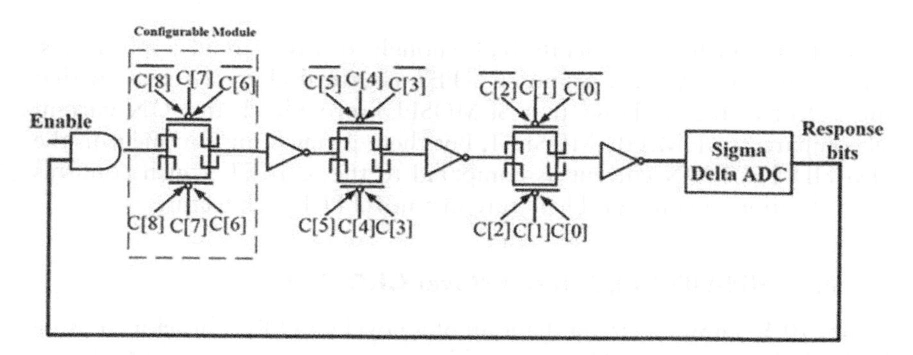

Figure 10.13 Proposed CRO PUF architecture.

Table 10.4 Truth Table of Configurable Module

PG0	*PG 1*	*CG*	*M1*		*M2*	
			Device polarity	on/off	Device polarity	on/off
0	0	0	n-HM-HD-MOSFET	off	p-HM-HD-MOSFET	on
0	0	1	n-HM-HD-MOSFET	on	p-HM-HD-MOSFET	off
0	1	0	n-HM-HD-TFET	off	p-HM-HD-TFET	on
0	1	1	n-HM-HD-TFET	on	p-HM-HD-TFET	off
1	0	0	p-HM-HD-TFET	on	n-HM-HD-TFET	off
1	0	1	p-HM-HD-TFET	off	n-HM-HD-TFET	on
1	1	0	p-HM-HD-MOSFET	on	n-HM-HD-MOSFET	off
1	1	1	p-HM-HD-MOSFET	off	n-HM-HD-MOSFET	on

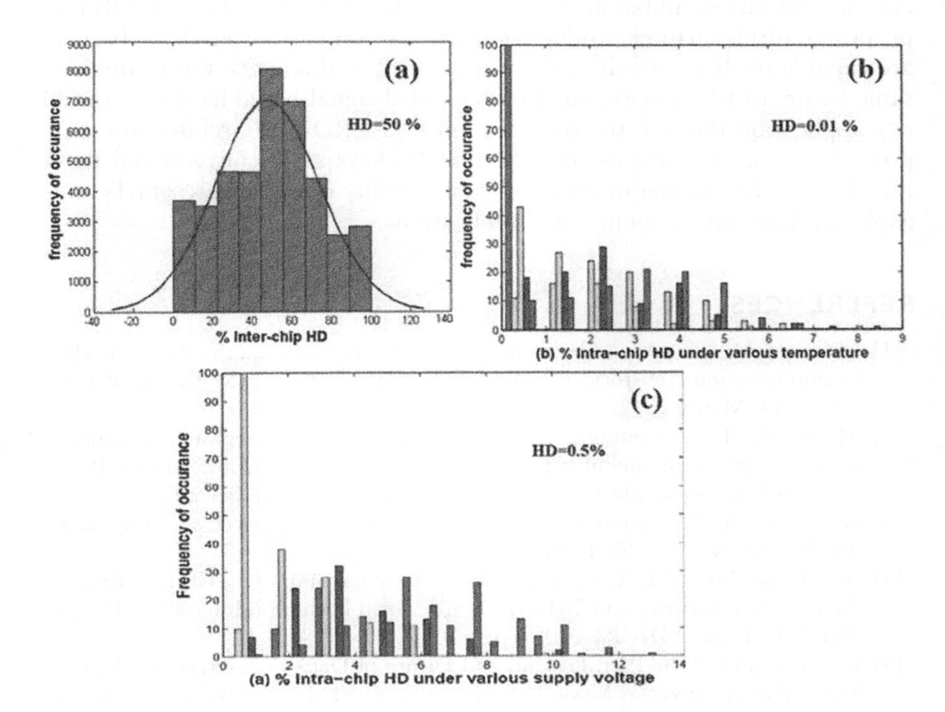

Figure 10.14 (a) Inter-HD, intra-HD with (b) varying temperature and (c) supply voltage.

are shown in Figure 10.14. Figure 10.14a shows that the inter-HD = 50% and that intra-HD on voltage variation is 0.5%, and for temperature, variation is 0.01%. Therefore, the proposed CRO-PUF produces a 2^N CRPs as compared to the 2N CRPs by CRO-PUF.

10.11 CONCLUSION

This chapter provides a detailed description of TFETs' working, advantages, drawbacks, existing solutions and applications. TFETs, which work on the inter-band tunneling principle and are not restricted by thermionic emission, have very low OFF current and SS compared to CMOSs. However, TFET has comparatively low ON current and high ambipolarity than MOS, which restricts its use in circuits. Besides, researchers had suggested many structural and physical modifications to upgrade TFET characteristics. In this sequence, we explored an HM-HD-RFET in a configurable PUF architecture. Basically, RFET is an electrically doped device that can be reconfigured either into n or p TFET or n or p MOSFET based on the potential applied over the external drain and source electrodes. Indeed, Si-based RFET produces satisfactory MOSFET results just like conventional one but lacks ON current and higher ambipolarity as in Si-TFET. Therefore, HM-HD-RFET proposed in this chapter produces sufficient drain current in all FETs that are capable to drive circuit applications. Also, in this work the reconfigurable nature of RFET is explored in the mixed-signal-based hardware security application through the construction of a CRO-PUF architecture that produces 2^N secure keys as compared to 2N keys as in conventional RO-PUF. Hence, this chapter opened a wide possibility to investigate TFETs and explored their utilization in novel applications.

REFERENCES

[1] Massimo, G., "The Age of Vacuum Tubes: Early Devices and the Rise of Radio Communications [Historical]", IEEE Industrial Electronics Magazine, Vol. 6, PP. 41–43, March 2012.

[2] Moore, G. E., "Cramming More Components onto Integrated Circuits". Available at: http://download.intel.com/museum/Moores Law/Articles Press Releases/Gordon Moore 1965 Article.pdf, Retrieved November 2006.

[3] Iwai, Hiroshi, "Roadmap for 22 nm and Beyond", Microelectronic Engineering, Vol. 86, No. 7–9, PP. 1520–1528, 2009.

[4] Iwai Idriss, Tarek, Haytham Idriss, and Magdy Bayoumi, "A PUF-Based Paradigm for IoT Security", in 2016 IEEE 3rd World Forum on Internet of Things (WF-IoT), IEEE, 2016, PP. 700–705.

[5] Robertson, D., "The Past, Present, and Future of Data Converters and Mixed Signal ICs: A Universal Model", Symposium on VLSI Circuits, PP. 1–4. IEEE, 2006.

[6] Chandrakasan, A.P., S. Sheng, and R.W. Brodersen, "Low-power CMOS Digital Design", IEEE Journal of Solid State Circcuits, Vol. 27, No. 4, PP. 473484, April 1992.

[7] Vasileios, A. Papaspiliotopoulos, George N. Korres, Vasilis A. Kleftakis, and Nikos D. Hatziargyriou, "Hardware-in-the-Loop Design and Optimal Setting of Adaptive Protection Schemes for Distribution Systems with Distributed Generation", IEEE Transactions on Power Delivery, Vol. 32, No. 1, 393–400, 2015.

[8] Idriss, Tarek, Haytham Idriss, and Magdy Bayoumi, "A PUF-Based Paradigm for IoT Security", in 2016 IEEE 3rd World Forum on Internet of Things (WF-IoT), IEEE, 2016, PP. 700–705.
[9] Boucart, K., and A.M. Ionescu, "Double Gate Tunnel FET with High-k Gate Dielectric", IEEE Transactions on Electron Devices, Vol. 54, No. 7, PP. 1725–1733, July 2007.
[10] Rajan, C., D. Sharma, and D.P. Samajdar, "Implementation of Physical Unclonable Functions Using Hetero Junction Based GAATFET", Superlattices and Microstructures, Vol. 126, PP. 72–82, 2019.
[11] Banerjee, S., W. Richardson, J. Coleman, and A. Chatterjee, "A New Three-Terminal Tunnel Device", IEEE Electron Device Letter, Vol. 8, No. 8, PP. 347–349, August 1987.
[12] ATLAS Device Simulation Software, Silvaco Int., Santa Clara, CA, USA, 2014.
[13] Iemys, Gravydas, Andrew Giebfried, Markus Becherer, Irina Eichwald, Doris Schmitt Landsiedel, and Stephan Breitkreutz-V. Gamm, "Modeling and Simulation of Nanomagnetic Logic with Cadence Virtuoso Using Verilog-A", Solid-State Electronics, Vol. 125, PP. 247–253, 2016.
[14] Higham, Desmond J., and Nicholas J. Higham, "MATLAB Guide Society for Industrial and Applied Mathematics", 2016.
[15] Ammar, Mahmoud, Giovanni Russello, and Bruno Crispo, "Internet of Things: A Survey on the Security of IoT Frameworks", Journal of Information Security and Applications, Vol. 38, PP. 8–27, 2018.
[16] Carbone, P., S. Kiaei, and F. Xu, Design, Modeling and Testing of Data Converters, Springer, Berlin, 2014.
[17] Pavan, S., R. Schreier, and GC. Temes, Understanding Delta-Sigma Data Converters, P John Wiley Sons, 24 January 2017.
[18] Robertson, and H. David, "Problems and Solutions: How Applications Drive Data Converters (and How Changing Data Converter Technology Influences System Architecture)", IEEE Solid State Circuits Magazine, No. 3, PP. 47–57, 2015.
[19] Narenda, S.G., "Challenges and Design Choices in Nanoscale CMOS", ACM Journal of Emerging Technologies in Computing Systems, Vol. 1, No. 1, PP. 7–49, 2005.
[20] Huang, Q., R. Jia, C. Chen, et al., "First Foundry Platform of Complementary Tunnel FETs in CMOS Baseline Technology for Ultralow-Power IoT Applications: Manufacturability, Variability and Technology Roadmap", IEEE International Electron Devices Meeting, PP. 22–2, 2015.
[21] Papaspiliotopoulos, Vasileios A., George N. Korres, Vasilis A. Kleftakis, and Nikos D. Hatziargyriou, "Hardware-in-the-Loop Design and Optimal Setting of Adaptive Protection Schemes for Distribution Systems with Distributed Generation", IEEE Transactions on Power Delivery, Vol. 32, No. 1, PP. 393–400, 2015.
[22] Halak, Basel, Mark Zwolinski, and M. Syafiq Mispan, "Overview of PUF Based Hardware Security Solutions for the Internet of Things", 2016 IEEE 59th International Midwest Symposium on Circuits and Systems (MWSCAS), PP. 1–4. IEEE, 2016.
[23] Rostami, M., M. Majzoobi, F. Koushanfar 1, D. Wallach 2, and S. Devadas, "Robust and Reverseengineering Resilient PUF Authentication and Key-Exchange

by Substring Matching", IEEE Transactions on Emerging Topics in Computing, Vol. 2, No. 1, March 2014.

[24] Rhrmair, U. et al., "PUF Modeling Attacks on Simulated and Silicon Data", IEEE Transactions on Information Forensics and Security, Vol. 8, No. 11, 2013.

[25] Huang, Q., R. Jia, C. Chen, H. Zhu, L. Guo, J. Wang, J. Wang, C. Wu, R. Wang, W. Bu, J. Kang, and W. Wang, "First Foundry Platform of Complementary Tunnel-FETs in CMOS Baseline Technology for Ultralow-Power IoT Applications: Manufacturability, Variability and Technology Roadmap", IEEE International Electron Devices Meeting (IEDM), 2015.

[26] Mutoh, Shinichiro, Takakuni Douseki, Yasuyuki Matsuya, Takahiro Aoki, Satoshi Shigematsu, and Junzo Yamada, "1-V Power Supply High-Speed Digital Circuit Technology with Multithreshold-Voltage CMOS", IEEE Journal of Solid-State circuits, Vol. 30, No. 8, PP. 847–854, 1995.

[27] Bazarjani, Seyfi, Lennart Mathe, Dana Yuan, Jeff Hinrichs, and Guoqing Miao, "Highvoltage Low-Power Analog Design in Nanometer CMOS Technologies", IEEE Bipolar/BiCMOS Circuits and Technology Meeting, PP. 149–154. IEEE, 2007.

[28] Zhu, Yan, Chi-Hang Chan, U-Fat Chio, Sai-Weng Sin, U. Seng-Pan, Rui Paulo Martins, and Franco Maloberti, "Split-SAR ADCs: Improved Linearity with Power and Speed Optimization", IEEE Transactions on Very Large Scale Integration (VLSI) Systems, Vol. 22, No. 2, PP. 372–383, 2013.

[29] Kester, Walt, "Which ADC Architecture Is Right for Your Application?", EDA Tech Forum, Vol. 2, No. 4, PP. 22–25, 2005.

[30] Chandan, B.V., K. Nigam, D. Sharma, and S. Pandey, "Impact of Interface Trap Charges on Dopingless Tunnel FET for Enhancement of Linearity Characteristics", Applied Physics A, Vol. 124, No. 7, PP. 503, 1 July 2018.

[31] Narang, R., M. Saxena, R.S. Gupta, and M. Gupta, "Linearity and Analog Performance Analysis of Double Gate Tunnel FET: Effect of Temperature and Gate Stack", International Journal of VLSI Design and Communication Systems, Vol. 2, No. 3, PP. 185, 1 September 2011.

[32] Zhu, Y., and M.K. Hudait, "Low-Power Tunnel Field Effect Transistors Using Mixed as and Sb Based Heterostructures", Superlattices and Microstructures, Vol. 2, No. 6, PP. 637–678, 2013.

[33] Lee, H., S. Park, Y. Lee, H. Nam, and C. Shin, "Random Variation Analysis and Variationaware Design of Symmetric Tunnel Field-Effect Transistor", IEEE Transactions on Electron Devices, Vol. 62, No. 6, June 2015.

[34] Holcomb, Daniel E., Wayne P. Burleson, and Kevin Fu, "Power-Up SRAM State as an Identifying Fingerprint and Source of True Random Numbers", IEEE Transactions on Computers, Vol. 58, No. 9, Septmber 2009.

[35] Jhaveri, Ritesh, Venkatagirish Nagavarapu, and Jason C.S. Woo, "Effect of Pocket Doping and Annealing Schemes on the Source-Pocket Tunnel Field-Effect Transistor", IEEE Transactions on Electron Devices, Vol. 58, No. 1, PP. 80–86, 2010.

[36] Saripalli, Vinay, Guangyu Sun, Asit Mishra, Yuan Xie, Suman Datta, and Vijaykrishnan Narayanan, "Exploiting Heterogeneity for Energy Efficiency in Chip Multiprocessors", IEEE Journal on Emerging and Selected Topics in Circuits and Systems, Vol. 1, No. 2, PP. 109–119, 2011.

[37] Kim, Minsuk, Youngin Jeon, Yoonjoong Kim, and Sangsig Kim, "Subthreshold Swing Characteristics of Nanowire Tunneling FETs with Variation in Gate Coverage and Channel Diameter", Current Applied Physics, Vol. 15, No. 7, PP. 780–783, 2015.

[38] Alam, K. "Orientation Engineering for Improved Performance of a Ge-Si Heterojunction Nanowire TFET", IEEE Transactions onElectron Devices, Vol. 64, No. 12, December 2017.

[39] Kumar, M.J., and S. Janardhanan, "Doping-Less Tunnel Field Effect Transistor: Design and Investigation", IEEE Transactions on Electron Devices, Vol. 60, No. 10, PP. 3285–3290, October 2013.

[40] Ghosh, B., and M.W. Akram, "Junctionless Tunnel Field Effect Transistor", IEEE Electron Device Letter, Vol. 34, No. 5, PP. 584–586, May 2013.

[41] Lahgere, A., C. Sahu, and J. Singh, "Electrically Doped Dynamically Configurable Field Effect Transistor for Low-Power and High—Applications", Electronics Letter, Vol. 51, No. 16, PP. 1284–1286, August 2015.

[42] De Marchi, M., D. Sacchetto, S. Frache, J. Zhang, P. Gaillardon, Y. Leblebici, and G. De Micheli, "Polarity Control in Double Gate, Gate-All-Around Vertically Stacked Silicon Nanowire FETs", IEDM, PP. 183–186, December 2012.

[43] Chithraja, Rajan, Jyoti Patel, Dheeraj Sharma, Amit Kumar Behera, Anil Lodhi, Alemienla Lemtur, and Dip Prakash Samajdar, "Implementation of P 4 ADC Using Electrically Doped III-V Ternary Alloy Semiconductor Nano-Wire TFET", Micro & Nano Letters, Vol. 15, No. 4, PP. 266–271, 2020.

[44] Chithraja, Rajan, and Dip Prakash Samajdar, "Design Principles for a Novel Lightweight Configurable PUF Using a Reconfigurable FET", IEEE Transaction on Electron Devices, Vol. 67, No. 12, PP. 5797–5803, 2020.

Chapter 11

Junctionless Transistors

Evolution and Prospects

Tika Ram Pokhrel and Alak Majumder

CONTENTS

11.1 INTRODUCTION

With the advent of technology, the trend of device scaling followed by Moore's law has reached to the sub-nanometer regime, where the device configuration of conventional transistors finds many adverse effects in their

DOI: 10.1201/9781003311379-11

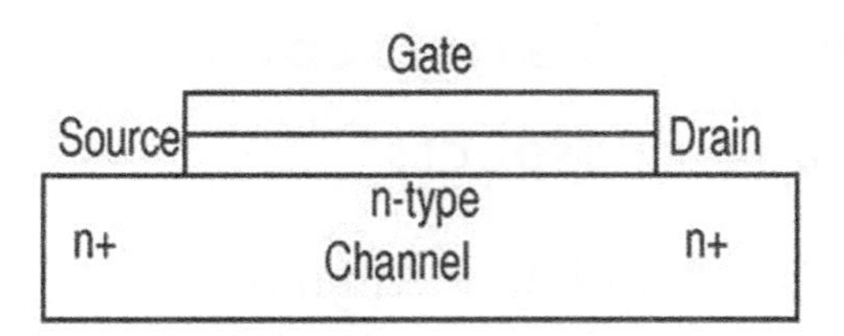

Figure 11.1 Two-dimensional schematic view of a simple JLT.

performance. When the channel length reaches a point that is equivalent to the size of the depletion region, the device becomes highly prone to short channel effects (SCEs). On the other end, a conventional transistor uses a semiconductor junction with the dopant present in the substrate. When the device reaches to the sub-nanometer regime, the junctions come very closer to each other, and it becomes very difficult to create a high potential gradient in the infinitesimal region that is needed for the carrier transport. This makes it very difficult to create a high-quality junction in these ultra-thin regions.

To address the previously stated issues, the junctionless transistor (JLT) is an admirable solution in this aspect to keep the trend of device scaling alive. The main feature of such transistors is that there are no PN junctions present, thereby eliminating the requirement of a high potential gradient. Accordingly, the fabrication process becomes less challenging even if a device of sub-10 nanometers is targeted. Since the JLTs have a uniformly doped and/or undoped channel, the device, by default, works in accumulation mode only. The simple structural idea of the JLT can be perceived from Figure 11.1. A few vital features of this device are strong immunity against SCEs, easy fabrication steps, lower cost, a lower electric field at ON state, steep subthreshold slope at very low drain biasing, and others. The channel must be very thin such that it can be fully depleted for the transistor to be in the OFF state when a JLT acts as a quasi-infinite resistance. However, to attain sufficient ON current, the channel must be heavily doped.

11.2 HISTORY OF JLTs

11.2.1 First JLTs

The idea of junctionless device was first framed by J. E. Lilienfeld in 1920, depicted in Figure 11.2 [1]. Due to the issues in the fabrication technology of ultra-thin devices, it took more than eighty years to fabricate physically the first JLT. The junctionless nanowire can be considered the first successfully modeled JLT fabricated at Tyndall Institute, Ireland, by Colinge et al. in

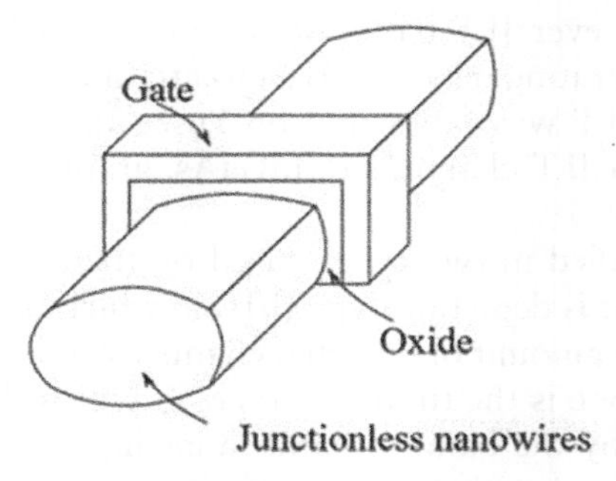

Figure 11.2 Structure of first junctionless nanowires [1].

2010. The name nanowire is due to the transistor with a channel length of less than 10nm, where the channel material is not monocrystalline but a thin film. The junctionless nanowires never operate in the inversion mode; rather, the characteristics of a JLT is more similar to the bulk MOSFET in the accumulation mode, when the channel is fully depleted. Since the evolution of JLTs, many other transistors of this kind (such as fin field-effect transistors [FinFETs], gate-all-around field-effect transistors [GAAFET], thin film) have been proposed and investigated. All these devices are characterized by junctionless structures, but they differ from each other considering the influence of technological parameters on their performance. FinFETs, GAAFETs, double-gated JLTs (DGJLTs) and tunnel field-effect transistors (TFETs) are a few JLTs that were fabricated in the last few decades. In multigate FETs, there exists an excellent coupling of the gate and channel that actually leads the device to deplete completely in the off state [1].

11.2.2 Evolution of Different Types of JLTs

After the successful fabrication and measurement of JLT in 2010, the doping concentration needed to be reduced to minimize the leakage current and the subthreshold swing (SS). The low temperature and high doping concentration resulted in an incomplete ionization of the carriers inside the channel, which degrades the ON current of the JLT. In the multigate FETs, there is a need to reduce the channel doping concentration in order to maintain a suitable threshold voltage and/or subthreshold slope. But it offers a penalty in terms of degradation of the ON current as the source to drain resistance increases. This issue is minimized by using the spacer in the source and drain region and by increasing the concentration in the source and the drain [1]. Furthermore, it was investigated that the electric field, perpendicular to the current flow is significantly low in a JLT, which is advantageous over bulk FETs for low-power applications.

Apart of the first-ever JLT based on silicon channel, there has been an evolution of incorporating new materials to attain higher mobility. The first germanium-based JLT was fabricated by Yu et al. [2]. The possible other materials used for a JLT channel are InGaAs, graphene, MoS_2 and $MoSe_2$ [3, 4], among others.

JLTs can be classified in two types based on the carrier transport inside the channel. The first is depletion-type JLTs, in which the amount of current flow depends on the amount of depleted channel steered by the applied gate voltage. The other one is the tunneling type, in which the electrical current is purely governed by the band-to-band tunneling of the carriers [5]. JLTs can also be categorized with respect to their geometric structure, material composition of the channel and the gate structure. It is known that the amount of current flow through the channel of any transistor is primarily modulated by gate biasing, and it highly affects the device performance. If the JLTs are designed like the conventional MOSFET with one gate terminal it is termed a single-gated JLTs, which has two configurations: bulk type and silicon-on-insulator (SOI) type. When more than one gate is involved in steering the channel current, it is termed a multigate JLT, which has many arrangements like DGJLTs, junctionless FinFETs and gate-all-around JLTs, among others [3]. These multiple gates may be controlled by a single-gate electrode or multiple-gate electrodes depending upon the application. It is comparably more advantageous in fabricating the multigate JLTs, because the involvement of many gates helps the device deplete completely. Since the speed of the device does not depend on the thickness of oxide, we can increase the capacitance value by adjusting the equivalent oxide thickness (EOT) to enhance the drain current.

11.3 CHARACTERISTICS OF JLTS

11.3.1 Conduction Mechanism

Unlike the bulk MOSFETs, which remain normally OFF and external gate biasing is required to turn ON the device, JLTs are generally an ON device. Due to the work function engineering between the metal and the semiconductor inside the channel and due to the nonexistence of junction, it will shift the threshold voltage and the flat-band voltage to some positive value [6] such that the flat-band voltage will be greater than the threshold voltage. When the device is in the flat band condition, the electric field perpendicular to the current flow is zero in the bulk channel. The variation of the drain current with the gate voltage in an inversion-mode MOSFET, a depletion-type MOSFET and a JLT are depicted in Figure 11.3 [5].

To determine the operating range of the device, the flat-band voltage and the threshold voltage are the most important parameters. The flat-band voltage has a direct impact on temperature, whereas the threshold voltage

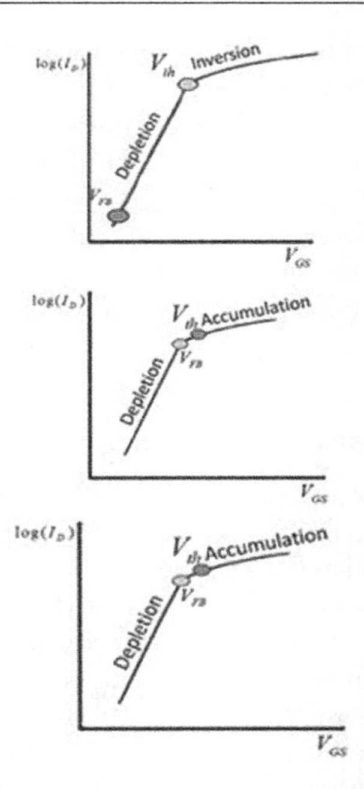

Figure 11.3 Drain current with gate voltage for (a) an inversion-mode MOSFET, (b) a depletion-type MOSFET and (c) a JLT [5].

decreases with the increase in the doping concentration. At threshold, JLTs have the peak carrier concentration in the channel being equal to the doping concentration. If the applied voltage is further increased beyond the threshold, the region of the depleted channel increases until the carrier concentration and the doping concentration holds equality with each other in the entire region. This leads to more enhancement in the drain current in a JLT as compared to a conventional MOSFET [6]. The thermal energy is not sufficiently high to ionize all the dopants at low temperatures and high doping concentration, thereby resulting in the increase in the source to drain channel series resistance, which extensively affects the drain current. The JLT also helps in minimizing the SS, drain-induced barrier lowering (DIBL), and the like.

The JLTs can be fabricated to yield low SS due to the impact-ionization effects, which can be obtained due to the high electric field when the device is operated in the saturation region. The high velocity of the carrier inside the channel in a saturation region guides to an increase in the channel potential, reducing the threshold voltage. But at the same time, the saturation current increases faster due to the formation of the carriers because of impact ionization. This process continues latching up the phenomenon and generates the carriers in an exponential way, leading to a steeper slope in the I_d–V_g curve. The region where the impact ionization takes place is comparably more in a JLT, and this guides the reduction of the drain-biasing voltage to obtain the sharp SS [6]. Therefore, it can also be concluded that the JLTs are essentially a low-power device. As the current in the JLTs flows effectively at the center of channel rather than in gate oxide interface(s), the chances of mobility degradation (due to scattering) are minimized, thereby enhancing the drain current, which is one of the advantages of the JLTs.

11.3.2 Region of Operation

As the JLTs can never be operated in the inversion region, we can term it a variable resistor, where the resistance varies horizontally from source to drain. Since there is no existence of the PN junction, the central region of the channel remains neutral and is entertained with high potential, which helps the device deplete more carriers from the source and drain regions. Since the carriers are present in these regions, the vertical electric field inside the channel within the flat-band region is zero. The thin channel in JLT allows the complete depletion of the channel at very low gate voltage. A better understanding may be realized from Figure 11.4, where the energy variation from source to drain is represented as a function of 'no gate biasing' and 'with gate biasing' [7].

In the flat-band region of operation and when the device is turned ON, the channel is nothing but a resistor with conductivity $\sigma = q\mu N_D$, where the mobility is of the bulk carriers in the channel and is rarely affected by the doping concentration.

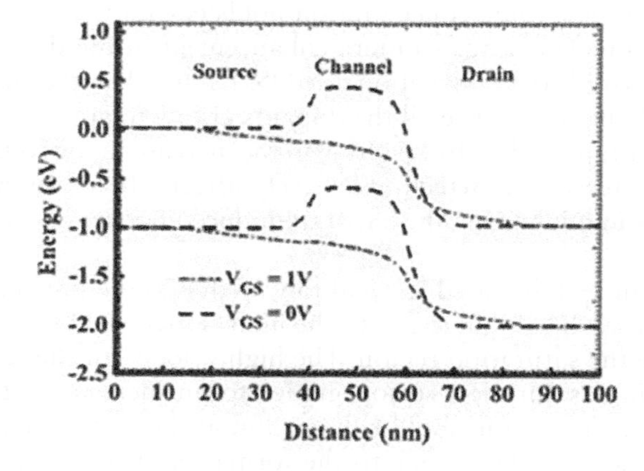

Figure 11.4 Lateral band diagram on JLT both on the ON and OFF states [7].

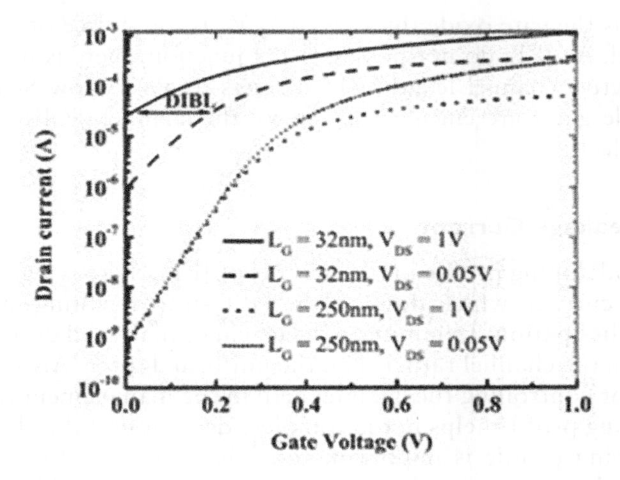

Figure 11.5 Short channel effect visualization for different channel lengths of a MOSFET [8].

11.3.3 SCEs

SCEs occur in devices where the channel length is comparable to the depletion-layer widths of the source and drain junctions, and that is never the case in long channel devices. As an example, we may refer to the I_D–V_G graph shown in Figure 11.5 [8], where the OFF current in the short channel devices increases very high compared to their long channel counterpart.

Likewise, down-scaling device technology corresponds to the reduction of threshold voltage, which results in a decrease in the slope of transfer characteristics (SS). All these factors are highly undesirable and called SCE, a few of which are mobility degradation, subthreshold current, DIBL, nonlinear threshold voltage variation, drain punchthrough and hot-carrier effects, among others. The presence of PN junction in the device leads to the reduction of the effective gate length, and hence, the effective gate length and the physical distance between the source and drain are not the same [9]. This also offers penalty in terms of SCE, which includes mainly DIBL, velocity saturation and hot-carrier effects [5]. The mathematical expression for the SCEs is expressed in Equation 11.1:

$$SCE = 0.64\frac{\varepsilon_{Si}}{\varepsilon_{ox}}[1+\frac{x_j^2}{L_{el}^2}]\frac{t_{ox}t_{dep}}{L_{el}^2}V_{bi} = 0.64\frac{\varepsilon_{Si}}{\varepsilon_{ox}}EI.V_{bi}, \tag{11.1}$$

Where ε_{Si} is the permittivity of silicon, ε_{ox} is the permittivity of gate oxide, L_{el} is the electrical (effective) channel length, x_j is the source and drain junction depth, t_{dep} is the penetration depth of the gate field in the channel region, t_{ox} is the gate oxide thickness and V_{bi} is the source or drain built-in potential. As JLTs do not possess a PN junction, there is no reduction of the effective channel length, which refers to a very low SCE. Also, if the multiple gates are engaged to control the channel, SCEs are further reduced [6].

11.3.4 Leakage Current

Due to good doping profile in the channel of JLTs, there is a good chance of leakage current, which flows in the OFF state, resulting in an undesired heat dissipation. To minimize the leakage, analytical doping may be exercised in the channel rather than the uniform doping. Also, the multiple gates for controlling the channel help in the management of leakage. A low doping profile helps in the complete depletion of the channel, and a high doping profile is important for enhancing the ON current [9]. Gate oxide leakage is another important aspect that highly affects the device performance. When the oxide thickness is very narrow, the carriers start tunneling towards the gate through the oxide layer and hence the use of the high-k oxide region is the potential alternative to address the same.

In the JLTs, due to narrow channel width and high doping profile, the band-to-band tunneling leakage has more impact. To further suppress this leakage, the bipolar JLTs can be considered [10]. In Figure 11.6, the comparison of the OFF current between bipolar JLTs and conventional JLTs is presented to note the superiority of bipolar JLTs.

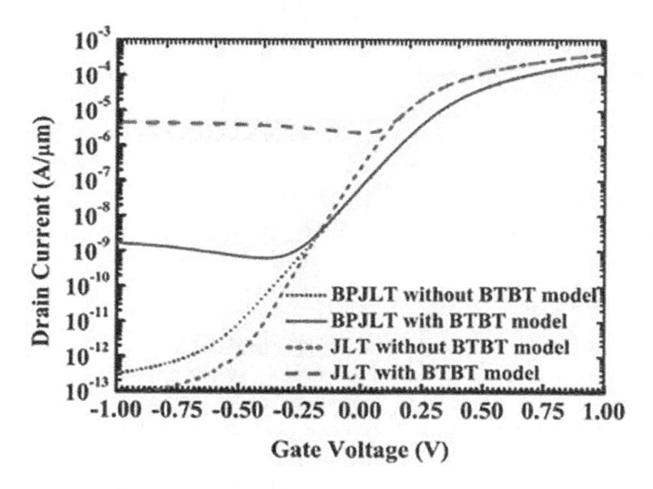

Figure 11.6 Transfer characteristics between conventional JLTs and bipolar JLTs [10].

11.4 MATHEMATICAL MODELING OF JLTs

11.4.1 Threshold Voltage Management

In today's era of low-power device reinvention, the study of the threshold voltage plays a vital role and is defined as the gate voltage that fully depletes (OFF state) the device layer. In the JLTs, the threshold voltage is mainly affected by doping concentration, silicon-film thickness, the width of the nanowire and gate oxide thickness. This dependency helps the circuit designer to create a certain threshold voltage for similar transistors in the same chip. The high doping profile in JLTs allows having a small threshold voltage, which may lead us to have more leakage current in the device.

With the trend of device scaling in the sub-nanometer regime, when the effect of SCEs comes into the picture the threshold voltage cannot be scaled down in a random manner, rather there is a definite mathematical expression to represent it as per Equation 11.2 [6]:

$$V_{th(SC)} = V_{th(LC)} - SCE - DIBL, \tag{11.2}$$

where $V_{th\,(SC)}$ is the threshold voltage for the short channel device and $V_{th\,(LC)}$ is the threshold voltage for the long channel device. This is called threshold voltage roll-off.

11.4.2 Current Equation Formulation in JLTs

In the JLTs, the bulk conduction is almost 100 percent unlike the conventional FETs. In this case, the modeling of drain current is discussed for the long-channel DGJLT with the parabolic potential approximation for the subthreshold and the linear region. The particular case considered here for the analysis is that both the gates are identical; that is, both metals have the same work function, and both front- and back-gate oxides are of same thickness. The first and basic current equation of the JLTs can be framed with the simplest structure of the JLTs and is expressed in Equation 11.3 [11]:

$$I_{Dsat} \approx \frac{1}{2}\frac{q\mu N_D W_{Si} T_{Si}}{L} V_{Dsat}^2. \tag{11.3}$$

Here,

$$V_{Dsat} = V_G - V_{FB} - \left(\frac{qN_D T_{Si}}{2\varepsilon_{Si}} + \frac{qN_D T_{Si}}{C_{ox}}\right). \tag{11.4}$$

From the current equation of a conventional MOSFET, we know that the drive current is directly proportional to the oxide capacitance, which is not the scenario in case of the JLTs, thereby making them faster than the conventional FET.

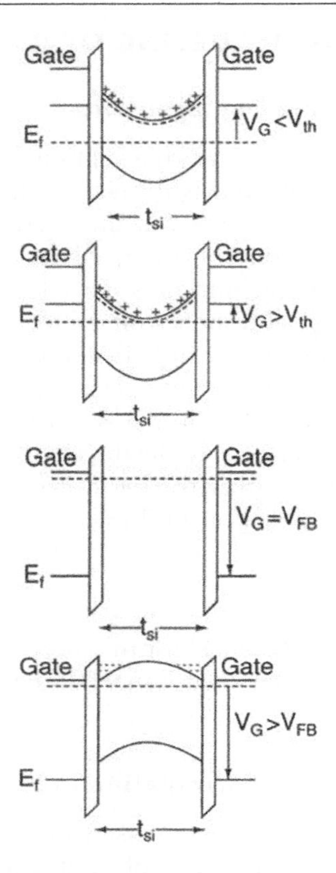

Figure 11.7 Potential variation from front to back gate in different biasing condition: (a) Subthreshold mode, (b) partially depleted bulk current mode, (c) flat-band mode and (d) accumulation mode [11].

The approximate work function difference between the metal and semiconductor is $E_g + q\nu T\ln(N_{si} / n_i)$. The channel is completely depleted, when the applied gate voltage (V_{GS}) is lesser than the threshold voltage (V_{TH}), and the bulk conduction mechanism is possible only if V_{GS}>V_{TH}. While moving from the front gate to the back gate, the potential is the minimum at the center of the device when the gate biasing is lower than the flat-band voltage [6]. As the device gets turned on, that is, the applied gate voltage is greater than the flat-band voltage, the central potential is greater than the potential present in the two interface regions. The potential variation for different gate biasing is depicted in Figure 11.7. The higher potential at the center of the channel leads to the depletion of the carriers from the source and drain regions as mentioned earlier. Figure 11.8 demonstrates the depletion of carriers from the source and the drain to form the channel as a function

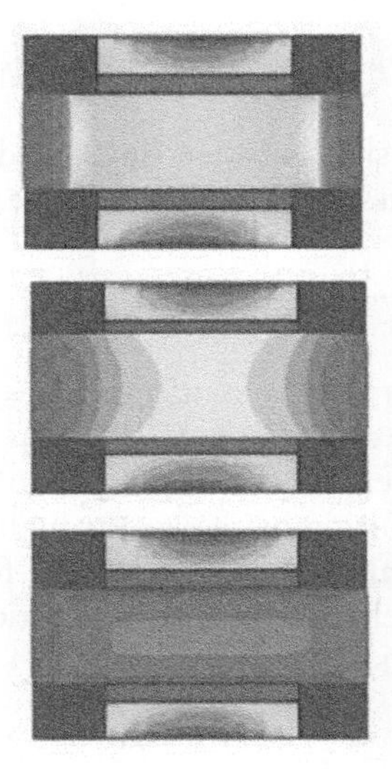

Figure 11.8 Depletion of the carried inside the channel from source and drain on different gate biasing: (a) *Vg* = 0.1, (b) *Vg* = 0.3 and (c) *Vg* = 0.6.

of applied gate biasing. It is perceived that any increase in gate biasing corresponds to more depletion from the source or the drain. At a certain voltage, there exists the complete depletion, and then the device is ON.

For an in-depth discussion on the current equation of JLTs, let us consider a simple DGJLT [11]. The poison equation can be reframed by considering only the mobile charges and the Pao–Sah gradual channel approximation:

$$\frac{d^2\varphi}{dx^2} = -\frac{qN_{si}}{\varepsilon_{si}}\left(1 - e^{(\varphi - V)\nu_T}\right), \tag{11.5}$$

where ϕ is the channel potential, ε_{si} is the permittivity of silicon and V is the electron quasi-Fermi potential.

By using the parabolic approximation (which is generally used only for subthreshold region) [8], the solution for the channel potential becomes

$$\varphi(x) = (4x^2 / t_{Si}^2) \times (\varphi_s - \varphi_o) + \varphi_o, \tag{11.6}$$

where φ_s is the surface potential and φ_0 is the potential at the center of the channel.

$$-\frac{\varepsilon_{ox}}{t_{ox}}(V_G - V_{FB} - \varphi_s) = -\varepsilon_{si}\frac{d\varphi}{dx}\Big|_{x=t_{si}/2} = \frac{4\varepsilon_{si}}{t_{si}}\Delta\varphi \tag{11.7}$$

$$\frac{d^2\varphi}{dx^2} = -\frac{qN_{si}}{\varepsilon_{si}}\left(1 - e^{(\varphi - V)\nu_T}\right) \tag{11.8}$$

Now, we need the relationship between the Ψ_s and Ψ_0. Applying the boundary condition and the Gauss's law at the interface, we have

$$\varphi(x) = \left(4x^2 / t_{si}^2\right) \times \left(\varphi_s - \varphi_o\right) + \varphi_o, \tag{11.9}$$

where E_{ox} is the permittivity of the oxide and $\Delta_\phi = \phi_o - \phi_s$ is the potential difference between surface and channel.

It is important to note here that the right part of Equation 11.7 is considered as the uniform distribution of charge throughout the channel such that $\Delta\phi \approx (tsi / 8\varepsilon si) \times (Q_{mobile} + qN_{Si}t_{si})$. The criterion for the threshold voltage for a fully depleted channel with the approximation that $\varphi_0 = 0$ is obtained from Equation 11.7; that is,

$$V_{TH} = V_{FB} - qN_{si}t_{si}t_{ox} / 2\varepsilon_{ox} - qN_{si}t_{si}^{\ 2} / 8\varepsilon_{si}. \tag{11.10}$$

Since the right part of Equation 11.7 only gives the information for half charge, it is necessary to involve more details on the total charge density. Therefore, one can get the total charge of the channel by integrating the charge density throughout its existence (inside the channel).

Using channel potential as shown in Equation 11.6,

$$Q_{total} = qN_{si}t_{si} - qN_{si}\int_{-t_{si}/2}^{t_{si}/2} e^{(\varphi - V)\nu_T}\,dx$$

$$= qN_{si}t_{si} \times \left[1 - \frac{e^{(\varphi - V)\nu_T}}{2}\sqrt{\frac{\pi\nu_T}{\Delta\varphi}} \times Erf\left(\sqrt{\frac{\Delta\varphi}{\nu_T}}\right)\right]. \tag{11.11}$$

Equation 11.11 is again not an analytical expression. However, $\Delta\varphi / V_T$ must be greater or equal to unity so that the error function tends to be unity. This condition is always satisfied, except when the applied gate voltage is approximately equal to V_{FB}. Hence, Equation 11.11 can be modified in addition to Equation 11.7 as

$$Q_{total} = 2\frac{4\varepsilon_{SI}}{t_{si}}\Delta\varphi = qN_{si}t_{si} \times \left(1 - \frac{e^{(\varphi - V)\nu_T}}{2}\sqrt{\frac{\pi\nu_T}{\Delta\varphi}}\right). \tag{11.12}$$

Figure 11.9 shows the pictorial representation of Equations 11.7 and 11.12 that describes the variation of the electric potential with respect to the position for different gate voltages [7]. It is clear that less gate voltage

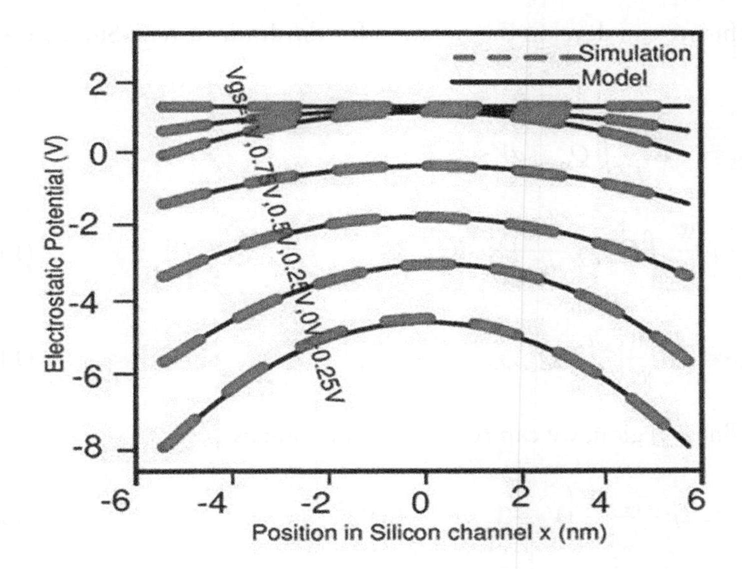

Figure 11.9 Electric channel potential (ϕ) distribution in the lateral position of the silicon channel for different gate voltages [11].

offers more bending in the potential profile, which is insignificant at higher gate voltage. Therefore, at low voltage (subthreshold regions), the potential difference is more, and at high gate voltage, the potential profile gets flattened. Hence, the potential difference $\Delta\varphi$ tends to be smaller. This also implies that φ_s and φ_0 are comparably the same above the subthreshold region.

Substituting $\Delta\phi = (tsi / 8\varepsilon si) \times (Q_{mobile} + qN_{Si}t_{si})$ and with some modification of Equations 11.7 and 11.12, we get

$$-\frac{2\varepsilon_{ox}\nu_T}{\beta t_{ox}}(V_G - V_{TH} - V) = Q_{mobile} - \frac{2\varepsilon_{ox}\nu_T}{\beta t_{ox}} \ln\left(-Q_{mobile}\sqrt{\frac{qN_{si}t_{si} + Q_{mobile}}{2\varepsilon_{si}\pi\nu_T q^2 N_{si}^2 t_{si}}}\right). \tag{11.13}$$

Here, $\beta = (1 + \varepsilon_{ox}t_{si} / 4\varepsilon_{si}t_{ox})$.

Since in subthreshold region, $Q_{mobile} << qN_{si}t_{si}$, we can rewrite Equation 11.13 as

$$-\frac{2\varepsilon_{ox}\nu_T}{\beta t_{ox}}(V_G - V_{TH} - V) = Q_{mobile} - \frac{2\varepsilon_{ox}\nu_T}{\beta t_{ox}} \ln\left(\frac{Q_{mobile}}{\sqrt{2\varepsilon_{si}\pi\nu_T q^2 N_{si}}}\right). \tag{11.14}$$

With this we can develop the equation for the drain current. Since we know that

$$I_{DS} = -\mu \frac{W}{L} \int_{0}^{V_{DS}} Q_{mobile} dV,$$

$$= -\mu \frac{W}{L} \int_{Q_S}^{Q_D} Q_{mobile} \frac{dV}{dQ_{mobile}} dQ_{mobile}, \tag{11.15}$$

$$I_{DS} = -\mu \frac{W}{L} \left(\frac{\beta t_{ox}}{4\varepsilon_{ox}} Q^2{}_{mobile} - \nu_T Q_{mobile} \right)\Bigg|_{Q_s}^{Q_D}. \tag{11.16}$$

In the linear region, we can reform the equation as

$$I_{DS} \approx 2\mu \frac{\varepsilon_{ox}}{\beta t_{ox}} \frac{W}{L} \left(V_G - V_{TH} - \frac{V_{DS}}{2} \right) V_{DS}. \tag{11.17}$$

On comparing Equation 11.17 with the current equation of the MOSFETs, we can say that the oxide capacitance, in the case of JLTs, gets reduced by a factor of β. This reduction is due to the existence of series capacitance formed due to the depletion of the channel [11]. The drain current in the linear region can be expressed as Equation 11.18:

$$I_{DS} \approx \mu \frac{W}{L} q N_{si} t_{si} V_{DS}. \tag{11.18}$$

It is to note that the drain current is directly proportional to the gate to source voltage, thereby guiding JLT to offer variable resistance as mentioned earlier. The device achieves the pinch-off region at the center, when the device operates in the saturation region [10, 12]. Hence, the saturation drain current equation can be represented as

$$I_{DS} \approx \mu \frac{\varepsilon_{ox}}{\beta t_{ox}} \frac{W}{L} \left[(V_G - V_{TH})^2 - \frac{\nu_T \beta t_{ox}}{\varepsilon_{ox}} \times \sqrt{q N_{si} \pi \nu_T 2\varepsilon_{si}} e^{(V_G - V_{TH} - V_{DS})/\nu_T} \right]. \tag{11.19}$$

Neglecting the exponential part from Equation 11.19, we have

$$I_{DS} \approx \mu \frac{\beta t_{ox}}{\varepsilon_{ox}} \frac{W}{L} (q N_{si} t_{si})^2. \tag{11.20}$$

Also, in the subthreshold region we can reframe the drain current equation for JLTs from Equations 11.14 and 11.16 and is expressed in Equation 11.21:

$$I_{DS} \approx \mu \frac{W}{L} \nu_T \sqrt{qN_{si}\pi\nu_T 2\varepsilon_{si} e^{(V_G - V_{TH})/\nu_T}} \left(1 - e^{-V_{DS}/\nu_T}\right). \tag{11.21}$$

This is the final expression, which is same as the subthreshold drain current computed in piece-wise model. The only difference is that the inversion current of DGJLTs is independent of the oxide thickness. Therefore, JLTs are more helpful in reducing SCEs by using the high-K gate oxide without affecting the inversion current. As JLTs never work in the inversion mode, the only modeling of the drain current is in the accumulation region, and it is presented in the following equations under different constraints [14].

Case I: $V_{GS} > V_{p0}$, $V_{GS} < V_{FB}$, $V_{DS} < V_{DSat1}$

$$I_D = \frac{q\mu_b N_D}{L_{effb}} \left(\frac{1}{n+1} \frac{S_{\max} - S_{\min}}{(V_{FB} - V_{p0})^n} \right. \\ \left. ((V_{GS} - V_{p0})^{n+1} - (V_{GS} - V_{DS} - V_{p0})^{n+1}) + S_{\min} V_{DS} \right) \tag{11.22}$$

Case II: $V_{GS} > V_{p0}$, $V_{GS} < V_{FB}$, $V_{DS} > V_{DSat1}$

$$I_D = \frac{q\mu_b N_D}{L_{effb}} \left(\frac{1}{n+1} \frac{S_{\max} - S_{\min}}{(V_{FB} - V_{p0})^n} (V_{GS} - V_{p0})^{n+1} + S_{\min} V_{DS} \right) \tag{11.23}$$

Case III: $V_{GS} > V_{FB}$, $V_{DS} < V_{DSat2}$

$$I_D = \frac{q\mu_b N_D}{L_{effb}} S_{\max} C_{ox} + \frac{\mu_{acc} C_{ox} W_{eff}}{L_{effacc}} \left(V_{DS}(V_{GS} - V_{FB}) - \frac{1}{2} V_{DS}^2 \right) \tag{11.24}$$

Case IV: $V_{GS} > V_{FB}$, $V_{DS} < V_{DSat1}$, $V_{DS} > V_{DSat2}$

$$I_D = \frac{q\mu_b N_D}{L_{effb}} \left[(S_{\max} - S_{\min})(V_{GS} - V_{FB}) + S_{\min} V_{DS} \right. \\ \left. + \frac{S_{\max} - S_{\min}}{n+1} \frac{((V_{FB} - V_{po})^{n+1} - ((V_{FB} - V_{DS} - V_{po})^{n+1}))}{(V_{FB} - V_{po})^n} \right] \\ + \frac{1}{2} \frac{\mu_{acc} C_{ox} W_{eff}}{L_{effacc}} (V_{GS} - V_{FB})^2 \tag{11.25}$$

Case V: $V_{GS} > V_{FB}$, $V_{DS} > V_{DSat1}$

$$I_D = \frac{q\mu_b N_D}{L_{effb}} \left[S_{\max}(V_{GS} - V_{FB}) + \frac{S_{\max} + nS_{\min}}{n+1} (V_{FB} - V_{po}) \right. \\ \left. + \frac{1}{2} \frac{\mu_{acc} C_{ox} W_{eff}}{L_{effacc}} (V_{GS} - V_{FB})^2 \right] \tag{11.26}$$

V_{po0}	: Linear pinch off voltage at $V_D = 0$
V_{p0}	: Pinch-off voltage ($V_{po} = V_{po0} - \eta V_{DS}$)
H	: DIBL coefficient
W_{eff}	: channel perimeter
S	: Nondepleted cross section of the channel
V_{Dsat1}	: Drain saturation voltage for the neutral bulk channel ($V_{Dsat1} = (V_{GS} - V_{po0})/(1 - \eta)$)
V_{Dsat2}	: Drain saturation voltage for the accumulation channel ($V_{Dsat2} = V_{GS} - V_{FB}$)
L_{eff}	: Effective length of the channel in neutral bulk channel
L_{effacc}	: Effective length of the channel in accumulation channel
μ_{acc}	: Mobility in accumulation mode
μ_b	: Bulk mobility

11.5 VARIOUS RESEARCH INVOLVING JLTs

11.5.1 Single-Gate JLTs

Single-gated JLTs consist of two types of device structures: one is the bulk well structure and the other is SOI. The schematic diagram of both configurations is shown in Figure 11.10. In 2011, the bulk planar JLT was proposed to have more current controllability over the SOI structure [15]. It is because the channel carriers can be tuned by two factors: doping and biasing the bulk well. Also, the OFF current can be minimized by supplying suitable body biasing. In the bulk JLTs, the effective channel thickness is always lesser than the physical thickness due to the built-in junction potential [16].

To sustain the single gate JLTs more in the sub-nanometer region, various techniques are exercised. For example, the work function engineering of the gate electrode has enhanced the I_{ON}/I_{OFF} ratio by 29% [17], whereas the differentially graded channel resulted in suppressing the I_{OFF} [18]. Using the high-k spacers increases the fringing of the electric field, which helps in improving the SS. However, this version of JLT is not significant enough to reduce the SCEs.

11.5.2 Double-Gate JLTs

Targeting the mitigation of SCE, many models of designing the double-gated JLTs have come up and the channel is sandwiched between two gates, which modulate the channel current as shown in Figure 11.11. There have been many modelings of this device carried out without considering the SCEs and quantum effects [19, 20], and few models are restricted to certain doping concentration only [21, 22]. Comparing with single-gated JLT, the device behavior is almost the same in the inversion region but is different in the accumulation and depletion region [5].

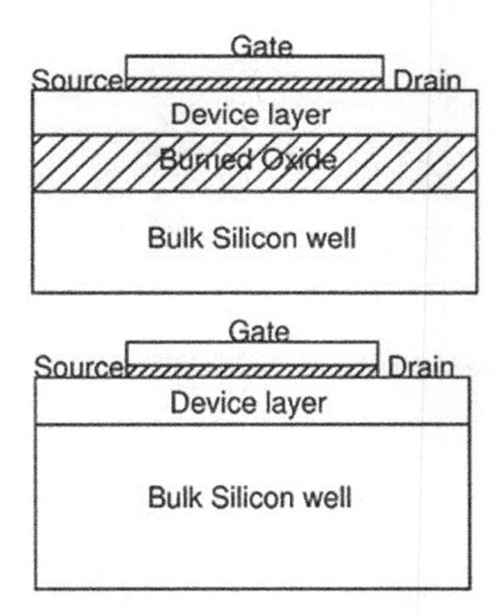

Figure 11.10 Single-gated JLTs: (a) SOI and (b) bulk well [15].

The high-profile doping in the device channel layer helps in the enhancement of the ON current along with the OFF current, which is not desired [23]. The requirement of threshold voltage increases with the rise in oxide thickness and device layer thickness [24] that help in the minimization of SS. The oxide thickness near the gate edges is increased in a few designs to minimize the leakage current. The spacer engineering on DGJLTs along with metal-gate work function engineering helps in enhancing the device performance as spacers have an influence on the lateral extension of the depletion width and, therefore, on the effective channel length [25, 26]. Like the single gate, the double gate JLTs offer lower leakage current in differentially graded channels.

11.5.3 FinFETs

FinFETs are one of the JLTs, whereby three gates are involved in controlling the channel and the device dimension strongly affects the performance. For example, an increase in the fin width of junctionless FinFETs from 6nm to 15nm can result in the variation of SS and DIBL by 42% and 60%, respectively. Likewise, the variation of channel length from 12nm to 21nm gives a variation of SS by 52% and DIBL by 14% [27]. It is important to note that increasing the fin height costs the analog performance extensively [28]. Similar to single gate JLTs, the junctionless FinFET also has two possible designs: bulk type and SOI structure. The SOI structure has more degree of freedom as compared to bulk structured junctionless FinFET [29]. Since the channel in FinFETs is modulated by agate on three sides as shown in Figure 11.12, it offers lower short channel effects as compared to the single-gate and double-gate devices [30]. The

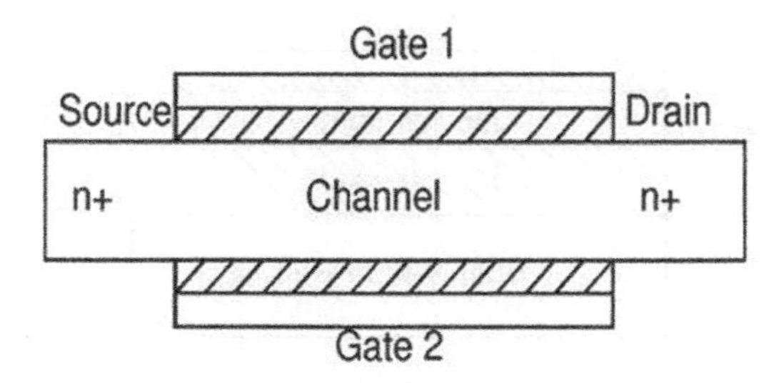

Figure 11.11 Schematic view of double-gated JLTs [5].

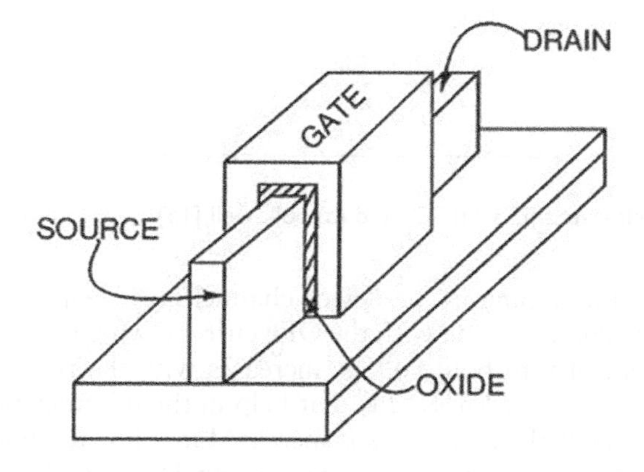

Figure 11.12 Structure of junctionless FinFET [32].

differentially graded channel junctionless FinFET offers an increase in ions by 21.1% [31, 32], and the threshold voltage is found to be very much sensitive to the metal-gate work function [33]. A lightly doped channel allows better gate control on the device [34]. The gate oxide engineering in FinFETs provides higher performance (in terms of Ion/Ioff and DIBL) after the implementation of complex hetero-gate oxide structures [35].

11.5.4 Nanowires

As stated earlier, the junctionless nanowire was first fabricated in 2010 and was one of the successful devices for ultra-thin fabrication techniques. One of the solitary behaviors of the nanowires is that the doping concentration of channel is the same as the source and/or the drain, but the gradient of the concentration is nearly zero. Therefore, no diffusion takes place and that helps in skipping the ultrafast annealing techniques in ultra-thin devices. Figure 11.13 depicts the generalized structure of junctionless nanowires, where the channel is very thin, helping deplete the channel completely in the OFF stage.

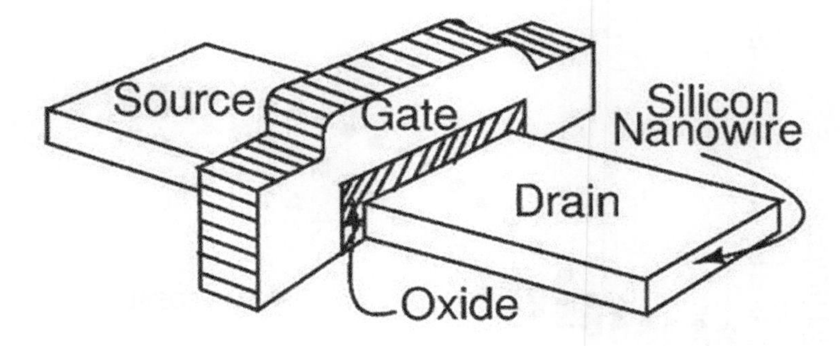

Figure 11.13 Junctionless nanowire [1].

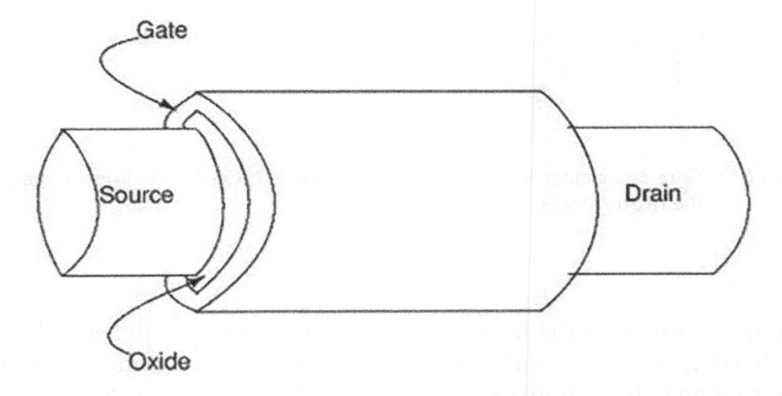

Figure 11.14 Junctionless xylindrical GAAFET [37].

11.5.5 GAAFETs

In GAAFETs, the channel is surrounded by the gate from all sides as depicted in Figure 11.14. The geometry of the channel determines the complexity of modeling the device behavior [35, 36] because of the involvement of cylindrical coordinates as compared to the rectangular-channel GAAFETs. However, the performance of cylindrical channel GAAFET is very high as the problem of corner effects is not present [37]. The speed of the device is inversely proportional to the thickness of the channel [38].

11.6 COMPARISON OF JLTS WITH DIFFERENT OTHER LOW-POWER FETs

The vertical electric field in JLTs is always less than in the case of inversion-mode or accumulation-mode MOSFETs. In the inversion-or depletion-type MOSFET, the carriers confined inside the inverted channel and

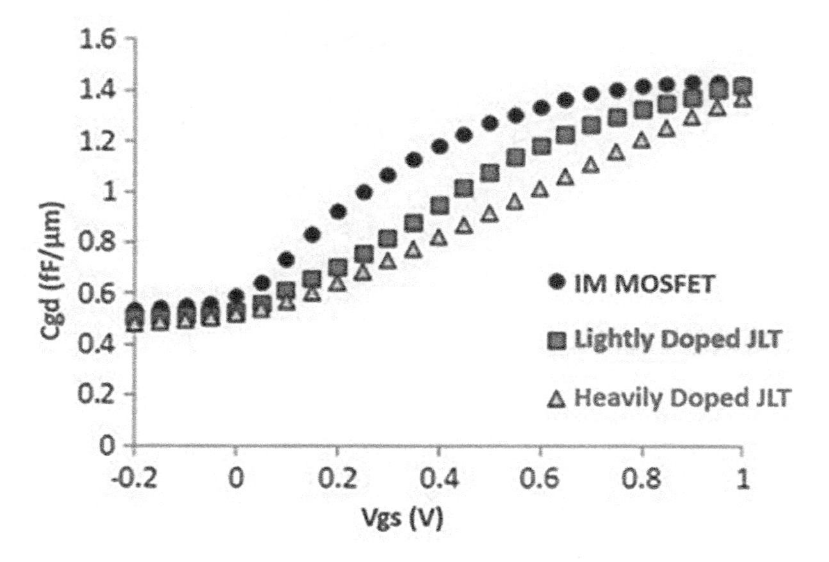

Figure 11.15 Gate capacitance for the inversion-mode bulk MOSFET, the lightly doped and the heavily doped channel JLT [42].

the doping profile in the inversion region are strongly influenced by the gate biasing (it is high only in the strong inversion region), where the scattering increases rapidly resulting to the degradation in the transconductance and the drive current. At the other end, the channel region is heavily doped in JLTs, and the carrier movement exists in the entire channel region for which the drive current is more than in conventional FETs [6].

The experimental data in Figure 11.15 depicts the gate capacitance for the inversion-mode bulk MOSFET, the lightly doped JLT and the heavily doped channel JLT [42]. It is found that the lightly doped JLT results in a lower gate capacitance than the MOSFET, followed by the heavily doped JLT. The lowering of this capacitance helps minimize the intrinsic device delay, and that is a huge advantage of JLTs.

Furthermore, the JLTs have the following advantages over conventional MOSFETs. First, it has as implified process flow as halo/extension that reflects in the removal of the deep source/drain implantation. Second, the implant activation anneal after gate stack formation is avoided, resulting in the reduction of the thermal budgets. Therefore, there is more flexibility in choosing the dielectric gate materials. The bipolar JLTs have the advantage of the full compatibility with the industry standard planar bulk twin-well CMOS process flow.

11.7 APPLICATIONS OF JLTS

Apart from the device level studies, it is very necessary to understand the time to market of any transistors. In other words, it is vital to realize the application of JLTs, which may become one of the backbones of the emerging electronic circuits. One of the applied areas is in the design of PUF for hardware logic security, which is heavily used in the applications of the Internet of Things [43]. A PUF produces true random numbers with high randomness and uniqueness, thereby making it difficult to replace or emulate by any other device and/or circuits and likewise enhances the security aspect. This helps protect the key encryption and decryption in the memory at any electronic gadget [44]. The main advantage of the JLT PUF (i.e., speed-optimized oscillator arbiter PUF) is the improvement in the frequency of generating the key. A multigate JLT can be used in the ultra-low power application devices. For the design of highly sensitive sensing applications, bipolar JLTs can be used. Also, a junctionless nanowire may be incorporated for low-power and high-frequency applications due to the heavily doped channel it possesses.

11.8 CONCLUSION AND FUTURE ASPECTS

JLTs, at the nanoscale regime, give good immunity to SCEs and produce higher Ion/I_{OFF} ratio. Also, the involvement of the high-k spacer helps reduce the leakage current massively. Although the band-to-band tunneling produces the minority charge carriers in the transistors, it is suppressed within the channel itself in case of the JLTs. The reduction of oxide thickness makes a JLT superior to a conventional FET in terms of speed. Hence, it may be considered a potential alternative for next-generation ultra-low-power VLSI design. The junction isolation in bipolar JLTs helps in the complete depletion of the channel in the OFF state.

Since the JLTs are of narrow width, the strained silicon technology may be used to configure the device to enhance performance. Work function engineering may also be implemented to the gate electrode for the maintenance of the threshold voltage.

REFERENCES

[1] Duffy, R., Tyndall National Institute, University College Cork, Lee Maltings, and Ireland Cork, "The (R)Evolution of The Junctionless Transistor", ECS Transactions, May 2016.

[2] Yu, R., Y.M. Georgiev, S. Das, R.G. Hobbs, I.M. Povey, N. Petkov, M. Shayesteh, D. O'Connell, J.D. Holmes, and R. Duffy, "Physica Status Solidi", Rapid Research Letters, Vol. 8, No. 65, 2014.

[3] Kim, Dong-Hyun, Tae Kyun Kim, Young Gwang Yoon, Byeong-Woon Hwang, Yang-Kyu Choi, Byung Jin Cho, and Seok-Hee Lee, "First Demonstration of

Ultra-Thin SiGe-Channel Junctionless Accumulation-Mode (JAM) Bulk Fin-FETs on Si Substrate with PN Junction-Isolation Scheme", Journal in Electronics Device Society. https://doi.org/10.1109/JEDS.2014.2326560.

[4] Peng, Bin, Wei Zheng, Jiantao Qin, and Wanli Zhang, "Two-Dimensional MX2 Semiconductors for Sub-5 nm Junctionless Field Effect Transistors", Journal of Materials (Basel), Vol. 11, No. 3, P. 430, Published online, 15 March 2018. https://doi.org/10.3390/ma11030430.

[5] Nowbahari, Arian, and Avisek Roy, "Luca Marchetti Junctionless Transistors: State-of-the-Art Electronics 2020, 9, 1174", 3 June 2020, Accepted: 17 July 2020; Published: 19 July 2020. https://doi.org/10.3390/electronics 9071174.

[6] Colinge, J. P., C. W. Lee, N. Dehdashti Akhavan, R. Yan, I. Ferain, P. Razavi, A. Kranti, and R. Yu, "Junctionless Transistors: Physics and Properties", Chapter February 2011. https://doi.org/10.1007/978-3-642-15868-1_10.

[7] Yanambaka, Venkata P., Saraju P. Mohanty, and Elias Kougianos, "Making Use of Manufacturing Process Variations: A Dopingless Transistor Based-PUF for Hardware-Assisted Security", IEEE Transactions on Semiconductor Manufacturing, Vol. 31, No. 2, PP. 285–294, 2018.

[8] Rios, R., A. Cappellani, M. Armstrong, A. Budrevich, H. Gomez, R. Pai, N. Rahhal-orabi, and K. Kuhn, "Comparison of Junctionless and Conventional Trigate Transistors with Lg Down to 26 nm", IEEE Electron Device Letters, Vol. 32, No. 9, September 2011.

[9] Gundapaneni, Suresh, "Investigation of Junction-Less Transistor (JLT) for CMOS Scaling", PhD Thesis, IIT Bomboy, 2012.

[10] Kranti, A., R. Yan, C.-W. Lee, I. Ferain, R. Yu, N. D. Akhavan, P. Razavi, and J.-P. Colinge, "Junctionless Nanowire Transistor (JNT): Properties and Design Guidelines", in Proc. ESSDERC, 14–16 September 2010, PP. 357–360.

[11] Juan Pablo Duarte University of California, Berkeley, "A Full-Range Drain Current Model for Double-Gate Junctionless Transistors", IEEE Transactions on Electron Devices, December 2011.

[12] Duarte, J.P., S.-J. Choi, D.-I. Moon, and Y.-K. Choi, "Simple Analytical Bulk Current Model for Long-Channel Double-Gate Junctionless Transistors", IEEE Electronics Device Letters, Vol. 32, No. 6, PP. 704–706, January 2011.

[13] Gundapaneni, Suresh, Swaroop Ganguly, and Anil Kottantharayil, "Bulk Plannner Junctionless Transistor (BPJLT): An Attractive Device Alternative for Scaling", IEEE Electronics Device Letters, Vol. 32, March, 2011.

[14] Colinge, Jean-Pierre, Tyndall National Institute, "University College Cork Junctionless Transistors" 978-1-4673-0836-6/12/$31.00 ©2012 IEEE.

[15] Jenifer, I., N. Vinodhkumar, and R. Srinivasan, "Optimization of Bulk Planar Junctionless Transistor Using Work Function, Device Layer Thickness and Channel Doping Concentration with OFF Current Constraint", in Proceedings of the 2016 Online International Conference on Green Engineering and Technologies (IC-GET), Coimbatore, India, 19 November 2016, PP. 1–3.

[16] Singh, D.K., P.K. Kumar, and M. Akram, "Investigation of Planar and Double-Gate Junctionless Transistors with Non-Uniform Doping", in Proceedings of the 2018 5th IEEE Uttar Pradesh Section International Conference on Electrical, Electronics and Computer Engineering (UPCON), Gorakhpur, India, 2–4 November 2018, PP. 1–5.

[17] Bora, N., P. Das, and R. Subadar, "An Analytical Universal Model for Symmetric Double Gate Junctionless Transistors", Journal of Nano- and Electronic Physics, Vol. 8, PP. 02002-1–02002-4, 2016.
[18] Yesayan, A., F. Prégaldiny, and J.M. Sallese, "Explicit Drain Current Model of Junctionless Double-Gate Field-Effect Transistors", Solid State Electronics, Vol. 89, PP. 134–138, 2013.
[19] Chen, Z., Y. Xiao, M. Tang, Y. Xiong, J. Huang, J. Li, X. Gu, and Y. Zhou, "Surface-Potential-Based Drain Current Model for Long-Channel Junctionless Double-Gate MOSFETs", IEEE Transactions on Electron Devices, Vol. 59, PP. 3292–3298, 2012.
[20] Cerdeira, A., M. Estrada, B. Iniguez, R. Trevisoli, R. Doria, M. de Souza, and M. Pavanello, "Charge-BasedContinuous Model for Long-Channel Symmetric Double-Gate Junctionless Transistors", Solid State Electronics, Vol. 85, PP. 59–63, 2013.
[21] Paz, B., F. Ávila-Herrera, A. Cerdeira, and M. Pavanello, "Double-Gate Junctionless Transistor Model Including Short-Channel Effects", Semiconductor Science and Technology, Vol. 30, PP. 055011, 2015.
[22] Chen, C.Y., J.T. Lin, and M.H. Chiang, "Threshold-Voltage Variability Analysis and Modeling for Junctionless Double-Gate Transistors", Microelectronics Reliability, Vol. 74, PP. 22–26, 2017.
[23] Wu, M., X. Jin, H.I. Kwon, R. Chuai, X. Liu, and J.H. Lee, "The Optimal Design of Junctionless Transistors with Double-Gate Structure for Reducing the Effect of Band-to-Band Tunneling", Journal of Semiconductor Technology and Science, Vol. 3, PP. 245–251, 2013.
[24] Mandia, A., and A. Rana, "Performance Enhancement of Double Gate Junctionless Transistor Using High-K Spacer and Models", in Proceedings of the 11th IRF International Conference, Delhi, India, Vol. 8, 25 May 2014, PP. 8–11.
[25] Gupta, M., and A. Kranti, "Sidewall Spacer Optimization for Steep Switching Junctionless Transistors Semiconductor", SciTechnol, Vol. 31, PP. 065017, 2016.
[26] Bharti, S. and G. Sensitivity Saini, "Analysis of Junctionless FinFET for Analog Applications", in Proceedings of the 2018 Second International Conference on Intelligent Computing and Control Systems (ICICCS), Madurai, India, 14–15 June 2018, PP. 1288–1293.
[27] Jegadheesan, V. and K.R.F. Sivasankaran, "Stability Performance of SOI Junctionless FinFET and Impact of Process Variation", Microelectronics Journal, Vol. 59, PP. 15–21, 2017.
[28] Rios, R., A. Cappellani, M. Armstrong, A. Budrevich, H. Gomez, R. Pai, N. Rahhal-Orabi, and K. Kuhn, "Comparison of Junctionless and Conventional Trigate Transistors with Lg Down to 26 nm", IEEE Electron Device Letters, Vol. 32, PP. 1170–1172, 2011.
[29] Rios, R., A. Cappellani, M. Armstrong, A. Budrevich, H. Gomez, R. Pai, N. Rahhal-Orabi, and K. Kuhn, "Comparison of Junctionless and Conventional Trigate Transistors with Lg Down to 26 nm", IEEE Electron Device Letters, Vol. 32, PP. 1170–1172, 2011.
[30] S. Xiong, and J. Bokor, "Sensitivity of Double-Gate and FinFETdevices to Process Variations", IEEE Transactions on Electron Devices, Vol. 50, No. 11, PP. 2255–2261, November 2003.

[31] Liu, F.Y., H.Z. Liu, B.W. Liu, and Y.F. Guo, "An Analytical Model for Nanowire Junctionless SOI FinFETs with Considering Three-Dimensional Coupling Effect", Chinese Physics B, Vol. 25, PP. 47305, 2016.
[32] Lü, W.F., and L. Dai, "Impact of Work-Function Variation on Analog Figures-of-Merits for High-K/Metal-Gate Junctionless FinFET and Gate-All-Around Nanowire MOSFET", Microelectronics Journal, Vol. 84, PP. 54–58, 2019.
[33] Tayal, S., A. Nandi, "Analog/RF Performance Analysis of Channel Engineered High-k Gate-Stack Based Junctionless Trigate-FinFET", Superlattices and Microstructures, Vol. 112, PP. 287–295, 2017.
[34] Sharma, M., M. Gupta, R. Narang, and M. Saxena, "Investigation of Gate All Around Junctionless Nanowire Transistor with Arbitrary Polygonal Cross Section", in Proceedings of the 2018 4th International Conference on Devices, Circuits and Systems (ICDCS), Coimbatore, India, 16–17 March 2018, PP. 159–163.
[35] Lime, F., F. Ávila-Herrera, A. Cerdeira, and B. Iñiguez, "A Compact Explicit DC Model for Short Channel Gate-All-Around Junctionless MOSFETs", Solid State Electron, Vol. 131, PP. 24–29, 2017.
[36] Moon, D.I., S.J. Choi, J.P. Duarte, and Y.K. Choi, "Investigation of Silicon Nanowire Gate-All-Around Junctionless Transistors Built on a Bulk Substrate", IEEE Transactions on Electron Devices, Vol. 60, PP. 1355–1360, 2013.
[37] Singh, P., N. Singh, J. Miao, W.T. Park, and D.L. Kwong, "Gate-All-Around Junctionless Nanowire MOSFET with Improved Low-Frequency Noise Behavior", IEEE Electron Device Letters, Vol. 32, PP. 1752–1754, 2011.
[38] Pratap, Y., R. Gautam, S. Haldar, R. Gupta, and M. Gupta, "Physics-Based Drain Current Modeling of Gate-All-Around Junctionless Nanowire Twin-Gate Transistor (JN-TGT) for Digital Applications", Journal of Computational Electronics, Vol. 15, PP. 492–501, 2016.
[39] Nobrega, R., Y. Fonseca, R.A. Costa, and U. Duarte, "Comparative Study on the Performance of Silicon and III-V Nanowire Gate-All-Around Field-Effect Transistors for Different Gate Oxides", in Proceedings of the XIII Workshop on Semiconductors and Micro & Nano Technology, São Bernardo do Campo, Brazil, 19–20 April 2018.
[40] Peng, Bin, Wei Zheng, Jiantao Qin, and Wanli Zhang, "Two-Dimensional MX2 Semiconductors for Sub-5 nm Junctionless Field Effect Transistors Materials (Basel)", Vol. 11, No. 3, PP. 430, Published online March 152018. https://doi.org/10.3390/ma11030430.
[41] Liu, Tung-Yu, Fu-Ming Pan, and Jeng-Tzong Sheu, "Characteristics of Gate-All-Around Junctionless Polysilicon Nanowire Transistors with Twin 20-nm Gates", Journal of Electronics Device, Digital Object Identifier. https://doi.org/10.1109/JEDS.2015.
[42] R. Rios, A., M. Cappellani, A. Armstrong, H. Budrevich, R. PaiGomez, N. Rahhal-Orabi, and K. Kuhn, "Comparison of Junctionless and Conventional Trigate Transistors with Lg Down to 26 nm", IEEE Electron Device Letters, Vol. 32, No. 9, September 2011.
[43] S. P. Mohanty, Nanoelectronic Mixed-Signal System Design. McGraw-Hill Educ., New York, NY, 2015.
[44] Xiong, S., and J. Bokor, "Sensitivity of Double-Gate and FinFET Devices to Process Variations", IEEE Transactions on Electron Devices, Vol. 50, No. 11, PP. 2255–2261, November 2003.

Index